# Modernizing American Land Records

# Order Upon Chaos

**Earl F. Epstein**
**Bernard J. Niemann Jr.**

*Cover image* Esri Map Book, Volume 25 *(Redlands, CA: Esri Press, 2010), 61; courtesy of FEMA Mitigation Directorate.*

Esri Press, 380 New York Street, Redlands, California 92373-8100
Copyright © 2014 Esri
All rights reserved. First edition 2014

Printed in the United States of America
18 17 16 15 14      1 2 3 4 5 6 7 8 9 10

*Library of Congress Cataloging-in-Publication Data*
Epstein, Earl F.
   Modernizing American land records / Earl F. Epstein and Bernard J. Niemann Jr.
      pages cm.
   Includes bibliographical references and index.
   ISBN 978-1-58948-304-0 (pbk. : alk. paper)—ISBN (invalid) 978-1-58948-375-0 (electronic)  1.  Land use—United States—History. 2.  Land use—
United States—Data processing. 3.  Land use—Information services 4.  Land tenure--United States. 5.  Land tenure—Government policy—United
States. 6.  Information storage and retrieval systems—Land use—United States. 7.  Geographic information systems—United States.  I.  Niemann,
Bernard J. II. Title.
   HD191.E77 2014
   025.06'33330973—dc23       2014008664

Ask for Esri Press titles at your local bookstore or order by calling 800-447-9778, or shop online at esri.com/esripress. Outside the United States,
contact your local Esri distributor or shop online at eurospanbookstore.com/esri.

Esri Press titles are distributed to the trade by the following:

*In North America:*
Ingram Publisher Services
Toll-free telephone: 800-648-3104
Toll-free fax: 800-838-1149
E-mail: customerservice@ingrampublisherservices.com

*In the United Kingdom, Europe, Middle East and Africa, Asia, and Australia:*
Eurospan Group
3 Henrietta Street
London WC2E 8LU
United Kingdom
Telephone: 44(0) 1767 604972
Fax: 44(0) 1767 601640
E-mail: eurospan@turpin-distribution.com

To the memory of Leon and Gertie Epstein, whose aspirations for a better life than that available to them in the Old World were realized, in part, through opportunities available to their son in America.

To the memory of Bernard and Emma Niemann, who individually left their German homeland for economic reasons after World War I, met in Chicago to pursue a new life, and imparted their aspirations to their son in America.

# Contents

# Preface

In 2007, the National Academies Press published *National Land Parcel Data: A Vision for the Future* (National Research Council [NRC] 2007). This report, the result of efforts by a panel assembled by the NRC, considered spatial information systems with a particular emphasis on land parcel data and land ownership. The 2007 panel referred to the NRC-sponsored report *Need for a Multipurpose Cadastre* (NRC 1980). The panel indicated that the 1980 report "became, and still is, a guidebook for land parcel data systems throughout the world" (NRC 2007, xi). This observation is striking because the 1980 report was written when the great changes in provision of spatial data made possible by geospatial and information technology were just beginning. The nature and scope of these changes could only be perceived, not observed, in 1980.

Another striking feature of the 2007 report was the observation that the 1980 report "advocated the development of a nationally integrated set of land parcel data and recommended a vision for achieving it. Yet 27 years later, despite technological advances that make it more feasible and policy directives that support the development of national land parcel data, the United States has not achieved this vision" (NRC 2007, xi).

The 2007 report stimulated us to consider and reconsider the bases for these observations, which resulted in this book. Earl Epstein was a member of the 1980 panel previously described. Both authors were members of other NRC-sponsored panels that addressed land parcel data and related activities. The authors have been university faculty in departments concerned with land and resource use (environment and natural resources for Epstein and both landscape architecture and urban and regional planning for Niemann). The efforts, or nature of the efforts, of these departments can be distinguished from those that emphasize *collection* and *supply* of land data and information. Rather, these departments aim to satisfy the *demand* for records and information in land planning and management. We believe that this experience brings an appropriate and specific perspective to consideration of the issues presented in this book.

Land records and information contain material about the location and status of land features. They also document the land rights, restrictions, and responsibilities (called *land interests* or *property rights*) associated with these features. Citizens, groups, organizations, and government officials desire knowledge of both the nature and extent of these land features and interests. Both aspects need to be known so that all are able to fully participate in the informed plans, decisions, and actions that determine use of land and its resources. Unfortunately, in the United States of America (referred to as *America* in most places in this book) the existing systems of land records and information operate in ways that do not satisfy this need for knowledge in the normal course of twenty-first-century American land planning and management. A particularly vexing problem is that material about the location of land features and interests, including maps, are not well connected to the rights, restrictions, and responsibilities attached to these features.

Evaluating these conditions generated the explanations and suggestions described in this book. A major reason can be summarized: American attitudes and practices regarding land concepts, land records systems, and land governance exhibit a strong, continuing preference for local-level actors and local government actions.

Institutions and processes at this level, including those that provide connections to data systems at higher levels of government, need more concentrated attention and organization if both local and national goals in land records and land management are to be achieved.

We propose a solution to American land data and information problems in the form of a modernized, multipurpose American land records system (ALRS). A major aspect of an ALRS that is not fully developed in the existing systems is a precise definition, identification, and use of authoritative data and information. These are established in the procedures and activities of the ubiquitous land planning and management process wherein land and resource uses are determined. Another aspect is recognition of and attention to American preferences for private and local government determinations of land and resource use. A third aspect is emphasis on the importance of parcels, parcel maps, and parcel identifiers as the basis for connecting location-based data with property rights records. A major concern is the feasibility of creating a national parcel database and map by connecting local parcel databases and maps.

This book encourages a dialogue about how to create workable partnerships in the effort to modernize American land records and information institutions. The objective of modernization is a system that satisfies the demand by all interested parties for data attributes and information that enhances their participation in land planning and management.

Geospatial and information technology have dramatically improved the provision and distribution of data about the location and status of observable land features. However, the technology has not been fully deployed to overcome the institutional barriers that inhibit connections between property records and location data, such as the American land planning and management process with its many formal, legal aspects. This land planning and management process provides an organizing context for the design and implementation of a modern, multipurpose ALRS.

Despite the frequently expressed individual and institutional demand for knowledge of both the nature and extent of interests in land and its resources, and despite the geospatial technology that has radically altered the ability to identify, measure, and map the location of land features, many land records remain in siloed information institutions, with organizations that accept all data without discrimination, or in no institution at all. The result is land records chaos, a major contributor to the recent mortgage crisis.

The continuing and growing complexity of twenty-first century American planning and management of its land and resources demands attention to the chaos. A land records and information system appropriate to the task of twenty-first century land planning and management is required. We believe the ALRS proposed in this book addresses this need. Land records and information are about more than provision of location-based services. They are also about who has the power to determine the use of land and its resources and how that power is allocated and exercised.

Part I of this book considers the nature and scope of existing American land records systems. It provides examples of problems generated in the normal course of land planning and management by the existing systems of land records and information. Part II relates the historical development of American attitudes and practices regarding land, land records, and land governance. This history is important because it shows that the problems cannot be solved by technical means alone. The history indicates both incentives for and barriers to change that are consistent with long-held American attitudes and practices. Part III describes achievable changes in American land records systems that are consistent with both these long-held attitudes and practices and the provision of material by geospatial technology. It presents action plans that ameliorate the chaos of American land records. The plans integrate the location-based material provided by geospatial technology with the demands of the land planning and management processes.

This book is for those concerned with how land and its resources are used, including those who develop geospatial technology and products, such as maps and databases, designed to serve those involved in land planning and management. The book also is for those who create land records and information, those who use these materials, and those who seek to bring these two groups together.

Specifically, this book is useful for the following individuals and groups:

- Elected and appointed officials concerned with or responsible for the activities of public land management agencies. Governors, mayors, county commissioners, and others are members of this group. Also included are legislators who give special attention to land-use issues among their many duties.

- Officials in public agencies with delegated authority for public land-use plans, decisions, and actions. These actors range from those who make approval and siting decisions for proposed energy-generating facilities in public utility agencies to those on local zoning boards who grant or deny a variance for a home business in an area that is zoned residential. Appropriate land records and information are needed to support these actions.

- Recorders, clerks, and assessors who collect, manage, and distribute land records and information.

- Landowners, real estate developers, bankers, utility companies, industrial developers, and others who have land interests and seek to sustain or alter land uses.

- Citizens and nonprofit groups interested in use of land and its resources in a community and motivated to sustain natural and cultural resources.

- Professionals, such as surveyors, lawyers, brokers, title insurers, land information companies, assessors, and others with recognized and long-established land- and land records–related activities.

- Geospatial technology professionals, including those involved in geographic and land information systems (GIS/LIS), who not only produce land data and the software to manage these data but also seek to better use their products in the determination of land and resource use.

- Educators in various land-related disciplines, such as geography, urban and regional planning, landscape architecture, natural resources, environmental studies, and public policy.

- Industry executives and staffs who provide the technical capacity and innovative solutions that support the activities of those listed above. These include industry and trade organizations that lead and set standards, publish best practices, and generally move industries forward in efforts to better manage and use land and its resources.

The panel that wrote the 2007 NRC report used the label "national land parcel data," whereas those who wrote the 1980 report used the label "multipurpose cadastre." *Cadastre* is not used often in America. The term is familiar to surveyors, officials in the US Bureau of Land Management (BLM), and parcel mapmakers. Even among some members of these groups, there is limited knowledge of the nature and scope of land records and information activities common to cadastres in other nations. Many Americans think of cadastres in terms of data about and maps of boundaries to parcels or other land interests, such as easements. Some reduce the concept of a cadastre to that of a parcel map. As a result, it has been hard to discuss the broader features of a cadastre or land records system, especially those aspects that include the rights, restrictions, and responsibilities that are attached to land features. Once, when one of the authors discussed a cadastre for an American audience, a listener said that a cadastre sounded like a cross between a cadaver and a disaster.

Notwithstanding the barriers to discussion and development of appropriate aspects of these systems, incentives for modernization of land records systems remain powerful. National and local goals regarding land records and information are compatible. The goal of citizen participation in land governance is particularly strong. However, more attention to local-level activities is needed because of particular American attitudes and practices. The details of these attitudes and practices, and of suggestions for change, are described in this book. The challenge today, as in 1980, remains as it was described by David Cowen in his preface to the 2007 NRC report: "How do we create workable partnerships to better serve our citizens?"

This book is about the common search for information that helps citizens and groups make informed determinations of land and resource use in twenty-first-century America. Current American attitudes and practices regarding land and resource use are part of American culture. The attitudes and practices are a significant part of the human dimensions of land planning and management. They influence how and why changes occur in the future. Knowledge of land history, land records, and land governance in America is essential to the search for the appropriate information, as much as is the search for the location of land features that geospatial technology can provide. Perhaps, this is the modern challenge to which the scientist, environmentalist, and writer Aldo Leopold referred when, in 1938, he wrote in *The River and Mother of God and Other Essays*, "Our tools are better than we are, and grow better faster than we do. They suffice to crack the atom, to command the tides. But they do not suffice for the oldest task in human history: to live on a piece of land without spoiling it."

# Acknowledgments

The nature and scope of American land records and information systems have a long history beginning before English settlement. These systems and their history are essential to understanding how and why land- and resource-use determinations have been, are, and will be made in America. These systems have two primary ingredients: maps of the location of land features and interests; and the substance of land rights, restrictions, and responsibilities. Knowledge of *both* aspects is essential to how American citizens choose to use their land and resources.

The authors have spent much of their professional lives concerned with American land planning and management. How to navigate environmental attitudes and practices has governed their attention to land records and information systems. This attention has led to questions of how geospatial and information technology serves the demand for data and information used in land planning and management. These questions are different from those of how to create and supply products. This approach is the conceptual focus for this book.

Both authors were strongly influenced, beginning in the 1970s, by the enthusiasm, dedication, wisdom, and support supplied so generously by James Clapp of the Civil and Environmental Engineering department of the University of Wisconsin–Madison and dean of engineering at the University of Maine. Land records and cadastral efforts in Madison were part of a dynamic, positive context established in the Institute for Environmental Studies (now the Nelson Institute) led by Reid Bryson and Arthur Sachs.

Ben Niemann acknowledges the inspiration and continual mentoring by Professors Charles Harris and Philip Lewis over these past five decades. Ben and his wife Sue wish to acknowledge and thank their many friends and colleagues at the Land Information of Computer Graphics Facility (LICGF) in the College of Agricultural and Life Sciences. They include Allen Miller, D. David Moyer, Nicholas Chrisman, Stephen Ventura, Jerome Sullivan, Allen Vonderhoe, David Tulloch, David Hart, Jane Licht, Math Heinzel, Thomas McClintock, Robert Gurda, Peter Thum, and the late Celeste Kirk and William Keenan who together helped foster, facilitate, and communicate the variety of the author's collective proof-of-concept experiments and solutions. The Niemanns also wish to thank the late Stephen Smith who, as the associate dean of the School of Natural Resources, helped in many ways to support their modernization ideas and facilitate their work. They wish to thank all the many students and visiting colleagues from around the globe who passed through the basement doors at LICGF and enriched their work. They wish to acknowledge two Wisconsin legislators of very different political persuasions, Robert Welch and Joseph Wineke, who fostered the passage of Act 31 and Act 39 in 1989–90. It was their courage and persuasive abilities that helped establish, fund, and sustain the Wisconsin Land Information Program (WLIP) over these past twenty-five years. The Niemanns also wish to identify the contributions of John Laub and the Wisconsin Power and Light Company who also helped convince two Wisconsin governors, Tony Earl and Tommy Thompson, to establish and fund the WLIP.

Earl Epstein learned how to conduct research by observation and collaboration with Larry Dahl, professor of chemistry at the University of Wisconsin–Madison and member of the National Academy of Sciences, and Ivan Bernal of Brookhaven National Laboratory. Professor Zigurds Zile of the University of Wisconsin–Madison Law School encouraged him to believe that it was possible and rational to move from chemistry toward law as a means of working on environmental issues. Unknown to him at the time, he was to be greatly influenced in the work described in this book by the seminal American legal historian J. Willard Hurst. Earl Epstein benefitted immeasurably from his colleagues in the exciting early years of the Civil and Surveying Engineering department at the University of Maine: Dave Tyler, Terry Keating, Alfred Leick, and Andre Frank. John Bossler of the National Geodetic Survey first supported and then continued to encourage work on land records and information. Gilbert Mitchell at the National Geodetic Survey was a constant friend and collaborator during two years of a professional leave with the agency. Earl Epstein's collaboration with Tom Duchesneau of the Economics Department at the University of Maine was a personal joy and professional pleasure. Craig Davis of the School of Environment and Natural Resources at The Ohio State University has been a constant friend and kindred spirit, as well as a professional colleague, throughout a quarter century in Columbus. Duane Marble was a friend in Columbus and provided advice and his wisdom in many professional matters. Colleagues and directors at the school made it possible for him to say that he could do what he wanted to do the way he wanted to do it.

One major influence on both authors, not adequately reflected in citations, is the work, personality, and spirit of John D. McLaughlin. James Clapp is reputed to have said in the early 1970s, in his inimitable way, "There's this kid from Canada, and he has the solution!" The graduate student we first knew in Madison grew up to be the president of the University of New Brunswick. North American cadastral and land records studies are enormously in his debt. This work, whatever its merits, would not be possible without him. His influence is felt on every page.

Brent Jones, global manager of cadastre and land records activities at Esri, read drafts of the manuscript, made specific and useful comments, and encouraged the authors. These efforts are greatly appreciated. We also thank the team at Esri Press for their assistance in the publication of this book.

The question arises in retrospect as to what conditions connected someone with both a PhD in chemistry and a law degree with someone who has two degrees in landscape architecture. Part of the answer is time and place. We met at the University of Wisconsin–Madison, renowned for actively "winnowing and sifting" ideas, in a state with a well-developed relationship between the university and the people known as the "Wisconsin Idea." However, long-term friendships and mutual respect for the other's contributions over three decades depend on more than an institutional setting. We are both children of immigrant parents and share both divergent and overlapping traditions and interests. Fortunately, we both graduated from superb high schools where high expectations are a given and optimism prevails. There is much yet to be done! We are very optimistic that the opportunity and benefits we suggest will emerge if we all collectively attend to the opportunity before us.

# About the authors

Earl F. Epstein (BS, chemical engineering; PhD, physical chemistry; JD, law) coauthored the National Academies of Sciences' 1980 report *Need for a Multipurpose Cadastre*. He taught quantum mechanics and survey law, and helped draft the social science component of the National Science Foundation's proposed Center for National Geographic Information and Analysis (NCGIA). Since 1988, he has studied and taught environmental, water, and natural resource law and policy at The Ohio State University, where he is professor emeritus, School of Environment and Natural Resources.

Bernard J. Niemann Jr. is professor emeritus, Department of Urban and Regional Planning, and Director of the Land Information and Computer Graphics Facility at the University of Wisconsin–Madison. He has degrees in landscape architecture from the University of Illinois (BFAALA) and the Graduate School of Design at Harvard University (MLA). For three decades, he taught geographic and land information systems (GIS/LIS) concepts and applications at both the undergraduate and graduate levels. In 1989, he helped establish and fund the Wisconsin Land Information Program (WLIP). He coauthored *Citizen Planners: Shaping Communities with Spatial Tools* (Esri Press 2010).

# Part I Introduction and problems

Part I describes how American land records and information systems fail to satisfy citizen demand for the knowledge needed to fully participate in twenty-first century American land planning and management. Chapter 1 presents a perspective on the nature and scope of the existing systems. It shows how geospatial technology, specifically geographic information systems (GIS), emphasizes the production of data about the location of land features but not the rights, restrictions, and responsibilities (called *property rights*) that individuals and institutions attach to these land features. The result is land records and information chaos for those who need both types of data and information in order to properly determine land and resource use.

Chapter 2 illustrates the problem by showing how inadequacies in the organization, connection, and provision of publicly and privately managed land records, encompassing both the nature and extent of land features and interests, hinder land-use planning and management. Chapter 2 shows how people and institutions create barriers to connecting records of both the location and nature of land features and interests.

These chapters encourage discussion of how disconnected land records are from land planning and management in America. The difficulty resides in the difference between aspects of land features that can be seen, measured, and located (e.g., parcel boundaries) and those aspects that are more abstract and unseen (e.g., the property rights that attach to a parcel). These chapters present ideas and practices designed to improve the ability of all citizens, groups, organizations, and officials to fully deploy geospatial technology and its products to improve their participation in land planning and management. These chapters also indicate that laws and legal processes that reflect and recognize preferred American attitudes and practices regarding land, land records, and land governance are the institutional part of the solution. A major objective of these chapters is to encourage a discussion of enhanced citizen participation in land planning and management based on a land records and information system that satisfies their demands. This demand is a powerful force for improvement of land records and information when combined with the supply of land data and information made possible by geospatial technology.

# Chapter 1
## Introduction

## 1.1   A land records and information perspective

In America it is difficult to ascertain the existence, location, status, availability, and connectedness of records and information on both land features and property rights. If this information is located, it often is difficult to assemble these two types of land records and information at the appropriate time and place and in an appropriate form for optimal land planning and management. This applies at the individual property parcel, community, and national levels.

Land records document much of what is known about land and its resources. Knowledge of all aspects of the land, from land features (e.g., location, size, and boundaries of the land) to property rights (e.g., rights, restrictions, and responsibilities placed on the land), is essential for wise land planning and management. Land records and information facilitate land planning and management activities, which include plans, decisions, and actions taken by citizens, groups, organizations, and officials who determine the use of land and its resources in a community.

A land records and information system, which houses land record documents and information, is distinguishable from other spatial information systems.[1] Although other systems also house data and information about the location and status of land features, land records and information systems include material about the nature of interests in land features (i.e., property rights).

American attitudes and practices, based on history, culture, economics, and law, keep these records and information separated and managed in independent institutions. Geospatial technology focuses on data concerning the location and status of land and its features on or near the earth's surface. The data often are collected and managed by agencies or subagencies that are not primarily involved in making land and resource determinations. Examples of these agencies include the local property assessor's office and the US Geological Survey. Other agencies, such as the register of deeds, receive, record, and document the results of negotiations and transactions concerning the allocation of property rights and boundary measurements. Further, records of privately arranged property rights of the type deposited in the register of deeds usually are segregated from records of publicly established property rights. The register of deeds does not always index documents in a manner that facilitates connecting these records and information. Land records and information systems in America are in a state of arrested development.

Each management institution has its independently established standards and practices, resulting in many public and private land-use determinations being made without an optimal package of timely, connected, and otherwise appropriate land data and information. Determinations made in this way often impose a variety of efficiency, effectiveness, and equity costs on people and the land.

Thinking about land records in the context of land planning and management requires considering how, by whom, and under what conditions land and its resources will be used in twenty-first-century America. This context is strongly influenced by the laws and the legal process that reflect the preferred attitudes and practices regarding land planning and management. Examination of these important laws and the legal process may reveal ways that land records and information systems can be changed to impose order on the chaotic world of land records and information.

Where content is concerned, land records are both more and less than the geospatial data and information commonly found in geographic, land, resource, and environmental information systems. Land records can contain *less* information than the vast array of geospatial data in geographic and other spatial information systems. For example, land records designated by law and legal process as authoritative support for land planning and management activities might not include all of the land data and information about the locations of land features that are available in various spatial information systems.

However, land records can contain *more* than the geospatial data and information in common geographic, land resource, environmental, and other location-oriented information systems in an important way. Geographic, land, resource, and environmental information systems focus on the location of land features on or near the earth's surface. The features can be natural—such as rivers, mountains, mineral deposits, and earthquake zones—or cultural, such as parcel boundaries, buildings, bridges, religiously or historically important sites, and five-star restaurants. The location of these features can be identified, and their positions can be measured and represented on maps. These data and information become the bases for the extraordinary array of location-based services that are now a part of the modern world. However, land planning and management are about more than the location of observable land features. The process of making plans and decisions and taking action regarding the use of land and its resources depends on who has the power to determine the use of land and its resources and how that power is exercised in a community. Each community has preferred attitudes and practices concerning land planning and management. Therefore, in addition to the location of observable land features, land records are about the unseen set of rights, restrictions, and responsibilities individuals and groups attach to land features—the property rights. The degree of attention to property rights records is what distinguishes land records systems from location-oriented systems.

## 1.2   A land records vision

A modern land records institution requires a vision. The vision described in this book is a land records and information system that allows all citizens interested in land and its resources to fully participate in land planning and management.

In this vision, the location-oriented results of geospatial technology would be better connected to records of property rights. Land interests generated in the land planning and management process would better direct the supply of geospatial technology products for modernization of land records systems.

The timeline for this vision extends to 2085, the two-hundredth anniversary of the Public Land Survey System (PLSS) in the United States. Devised in the 1780s, the PLSS encouraged land development by means of its divestiture by the federal government into private hands, security of boundary and land interests for those who received land, provision of preliminary indication of land conditions by government agents for potential developers, and support for local government and public education.[2] The PLSS successfully managed the difficult task of combining the scientific and technical challenges of land measurement with development of property rights consistent with American attitudes and practices. The vision and objectives for the PLSS are appropriate for a modern American land records system even if modern land-use challenges are more complex than in the eighteenth century. For those areas where the PLSS was not established, lessons from its history and current practice remain informative and valuable for land records and information system development.

Modernization of land records requires thinking institutionally (Heclo 2008). Institutional thinking means not limiting thought to a particular institution. A review of history and practice reveals that many systems operate as independent institutions, which encourages thinking within the institutions rather than across them. An institutional perspective asks the various actors in the independent institutions to adjust and balance their self-interests and actions to achieve a common, beneficial objective.

A land records and information institution must not only provide the location of land features but also identify land features' rights, restrictions, and responsibilities. The institution must connect this information at the time, place, and manner appropriate to the demands of all those who participate in land planning and management. In twenty-first-century America, this coordination is important to those who emphasize public as well as private interests because land has become "a commodity affected by a public interest" (Babcock and Feurer 1979, v). In a democratic society that has a goal of citizen participation in governance of land and resource use, all citizens must have an efficient, effective, and equitable role in the land planning and management process. Each citizen must seek to use land data and information to fully participate in the process. Citizen engagement and empowerment can be harnessed as an external force for overcoming the barriers to thinking institutionally.

For citizens to fully participate in land governance, they need access to a "two-way information portal" that enables them to do the following:

- submit records and information to a receptive land records and information agency

- acquire material from an appropriately open agency that uses the material in the normal course of its duties

This two-way information portal is part of the vision for a modern American land records and information institution.

A modern land records system constructed in this way would operate in a balanced manner for citizens, groups, and organizations in the contexts of community, market, and state in determining the best use of land and its resources. A balanced set of interests would establish a symbiosis between actors in a community, providing for informed citizens with connections to others.

The ideas of citizen participation and empowerment would have the additional advantage of appealing to many in the community, which could be harnessed to gain and sustain support for modernization.

## 1.3    Land use and land records

Land-use activities and land records are connected by the need to document the status of the land itself and the status of ownership. Both factors are important to those who seek to efficiently, effectively, and equitably determine how land and its resources are used. The status of land and ownership are connected by geospatial technology and the art and design of land planning and management.

Geospatial technology greatly enhances our ability to collect, manage, analyze, characterize, represent, and distribute data and information about the location and extent of features on or near the earth's surface. The technology's development over recent decades encompasses remarkable, perhaps revolutionary, change in both land measurement and management of the resulting observations.

The location and extent of features on or near the earth's surface can be measured more accurately, more rapidly, and less expensively than was imaginable a few decades ago. These observations and measurements can be stored, combined, characterized, represented, and distributed in and among computers in ways that were not possible, practical, or economical in the recent past. Many of these tasks can now be done easily and quickly, and new tasks can be undertaken.

Geospatial technology significantly increases the number and diversity of individuals and institutions who can use or acquire the results of the technology and land records. In the past, the tools and expertise were available only to a few with wealth or expertise, but with the technological advances, many more interested citizens have access to its use and products, enabling increased participation in modern American land planning and management.

The art and design of land planning and management encompass attitudes and practices that are important to community members. These important cultural values and behaviors are crucial factors that determine land-use plans, decisions, and actions.

Cultural values and behaviors are reflected in land records and information. The most important set of information is the established land interests in a community—the recognized rights, restrictions, and responsibilities attached to land features, especially parcels. Land interests determine what can be done with a parcel or area, which parties have the power to decide, and how that power is exercised. These land interests, called *property rights*, are established by both public and private activity in the American legal system.

These important land interests are documented in land records. Records of these property rights constitute a distinctive set of land records often separated from the set of location-oriented records that are the common products of geospatial technology.

Controversy often exists over what constitutes appropriate land use, the balance between private and public determination of that use, the nature and extent of individual and collective participation in land-use governance, and investments in land development and sustenance. Controversy also exists over the extent of public access to land records and information in government offices, investments in land records, the distribution of land data, and who is empowered by the data and technology. The controversy over access is reflected in continuing litigation over the meaning of the federal Freedom of Information Act (FOIA) and state open records laws. These controversies and resulting uncertainties over access to, investments in, and distribution of land records lead to the collection, management, and sale of public land records by private organizations, such as title and mapping companies.

The nature and scope of coordination between the land records associated with the location of land features and those associated with land interests is not a technical problem alone. This coordination involves existing land records institutions whose structure determine who, when, where, and under what conditions interested parties can acquire and use land data and information. These are cultural issues involving the distribution of information control in a community, which contributes significantly to the determination of how land and resources are used.

## 1.4    Scope and context for a modern American land records and information system

Land planning and management are important and specific to each community. Community members have individual and collective attitudes and practices regarding land and resource use. When community members form a preferred set of standards regarding this use, these standards determine who has the recognized power to decide land and resource use and how that power is exercised. When these preferred attitudes and practices are channeled and formalized in law and legal process, they constitute a land- and resource-use institution—a land planning and management process.

Land management and planning encompass a variety of land design activities on different scales. One scale involves actions of individuals or organizations on their property.

A second scale involves the landscape design for a site that includes several or many parcels. A third scale includes the work of local planning or other agencies with responsibilities for large areas of land. The work of agencies—such as the federal Bureau of Land Management, US Forest Service, or National Park Service—encompasses the whole nation.

Design activities include gathering appropriate data and information about the nature and extent of land features and interests. These data and information are of concern before, during, and after parcel, site, or area design or land-use change.

Everyone is interested in or affected by how land and its resources are used. They care deeply about what they can do with their land, what other individuals can do that affects land, and what is done collectively by the community that influences land use. These concerns may be expressed in the form of specific questions, such as the following, which embody many of the fundamental issues confronted in land planning and management:

- How likely is it that my land will flood?

- How well protected is a nearby wetland?

- What data about the location of a land feature can I obtain and present to a receptive official in an administrative process designed to protect the resource?

- Who and where is the nonresident owner of an urban property to whom a building code violation needs to be served?

- What is the likely effect of a proposed zoning change on my land?

- What is the likelihood that mortgagors in a bundled mortgage instrument will continue to pay?

- What can I do and not do with my land?

- Can I build a shed at the back of my property?

- If I create a land trust or an environmental easement now, what is the likelihood of its survival?

- Can I cut trees on my land that are along a designated wild and scenic river?

People seek answers to these questions in order to take action. People often lack the confidence to act because they do not know enough about the status of land and its features or about who has the power to determine their use.

Many of these questions can be answered but at considerable expense, time, and effort. Some have the resources to identify, locate, gather, analyze, characterize, and distribute the land records and information that answer many of these questions, and some do not have the resources. This difference significantly affects who determines land and resource use, how actions are determined, who benefits, and the quality of the resulting environment.

Many private-sector sources supply nonauthoritative land information (e.g., the Zillow Corporation). The Internet makes these sources easily accessible. People looking for reliable and timely answers will use any data and information they can get, but many users do not know what material is most appropriate for their intended use. With the massive amounts of data available, some users are under the illusion that all necessary data are available. Users want to reduce uncertainty to the point that they feel sufficiently secure in their repose or in knowledge of the results of a proposed land-use change by their neighbor or the whole community. Not everyone seeks certainty. Many live by the principle "reduce uncertainty to where you can absorb the residual uncertainty and act."

It is reasonable to assume that plans, decisions, and actions regarding land and resource use and management will become more complex during the twenty-first century. Population growth, fundamental changes in the environment, changes in the nature and location of human activity, and desire to preserve traditional practices, not to mention the increased strain on limited natural resources, ensure this added complexity. Increased complexity means more actors, increased numbers and types of land concerns, increased numbers and types of interactions between people regarding land use, and a demand for appropriate data and information to influence informed actions.

## 1.5    Actors and their roles

Law and legal process establish recognized rules and procedures for public land planning and management. The rules and procedures are the result of statutes, such as zoning ordinances or ordinances controlling use of a natural resource, delegating authority to an agency to make rules and grant permits for proposed changes in land use. The rules and procedures define how an agency executes its delegated or authorized duties. A framework is established for the activities and for the role of the actors in the public process. The framework also substantially influences the behavior of actors in their private actions.

Laws and legal process also define the rules and procedures for the related land records systems. One set of rules governs the documentation of privately arranged creation and transfer of land interests. Examples of these statutes, rules, and procedures include property, real estate, mortgage, and other substantive laws that define property interests. Another set of rules governs how public agencies establish and document public land-use controls and similar actions that affect land interests. These include zoning, environmental, and land-use-control laws. A third set of rules governs the related, often symbiotic, actions of both public and private actors involved with land records and information. These statutes, rules, and procedures cover title recordation, title insurance, surveys, and so on. They also include federal and state administrative procedures acts, freedom of information and open records acts, and environmental policy acts. Together, these public and private laws make formal the attitudes and practices that a community prefers regarding the private and public actions that determine use of land and its resources and the management of records that document the results of land planning and management.

In the modern world of geospatial and information technology, actors in the land planning and management process expect to find answers to their questions about the physical status of land. They also have a reasonable expectation of finding information about the status of rights, restrictions, and responsibilities regarding the use of land and its resources.

The major classes of actors in the public land planning and management process include those who propose to change the land use, those officials who have the legally delegated authority to approve the proposal, and those who feel that they will be affected by the result. These actors include landowners; citizens; citizen groups; private-sector organizations; elected officials; agency officials; land developers; nonprofit groups; land records and information managers; land measurement, geospatial, and information scientists and engineers; and educators.

Activities in the public venue wherein the land planning and management decisions are made have been called a "land-use drama."[3] The relation between the major classes of actors in this drama is shown in figure 1.1.

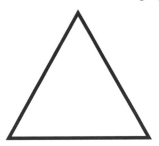

Applicant for a land use change permit

Agency with the power to decide          Other interested parties

**Figure 1.1** Actors in the public land-use drama.

Law and legal process establish the venue, procedures, and standards for use of land records and information presented to agencies. The rules and procedures are specific to the community where the decisions are made.

The actors seek land data and information so that they can actively, forcefully, and fully participate in the process. If those affected by changes in land uses do not accept the data presented or available to them, they can dispute the data's validity. Land data and information presented by agency officials or others as the informational basis for a land-use determination decision can be challenged and found wanting in the law-based process. Agency officials who have the delegated power to make decisions can reject the material and look elsewhere for appropriate land data and information. This contentious activity is typical of the land planning and management process. It provides a message to all who participate in the process.

"An adversarial relationship can exist wherever data is a matter of dispute. The idea that a neutral agency will create a noncontentious data set for decision making is inconsistent with our adversarial legal system and is not good policy in a society where we respect the right of the individual to question the facts that are used to determine rights and interests."[4] Messages such as these from regular participants in the land planning and management process indicate that land-feature-location records and land-interest records are not adequately connected. The messages also indicate that producers and users of land records and information are not fully coordinated in the American land planning and management process. The push of geospatial technology is not adequately connected to the pull of land planning and management.

The challenge for the actors is to design a land records and information institution that closes the gap between the land records that are supplied and the land records that are used in the process of making plans, decisions, and actions regarding land use. Those who use geospatial technology to supply land data, and the maps and other forms of information that represent these data, emphasize locations of land features. Those who use land records and information seek material that effectively expresses their attitudes.

Examination of the public land planning and management process, the role of actors in that process, the land records and information used in the process, and the messages from that process form a basis for a method designed to close the gap between land records that are supplied and those that are actually used.

# 1.6   Conclusion

Modernization of the land records and information institution requires a focus on the land planning and management process. This process provides a venue for citizens and organizations where the location-based products of geospatial technology are connected to documentation of land interests in the determination of land and resource use.

Social and cultural attitudes and practices, especially those related to knowledge of both the nature and extent of land interests, are a major domain of inadequately addressed problems in the development of geospatial technology for land planning and management.

Modernization of land records requires an answer to the following questions: What land records and information system serves the needs of all citizens in the effort to best plan and manage the use of land and its resources in twenty-first-century America? What practical and acceptable changes will move the current system toward this ideal system?

Mechanisms for modernization in the American land records and information system are available and consistent with the potential offered by both new geospatial technologies and the nature of modern land planning and management. American law and legal process associated with land planning and management provide a rich source of observations and messages that can be used in the design of a modern land records and information system.

Chapter 2 contains examples of land planning and management problems that arise because of the common attributes of land records and information in and available from existing American land records and information systems. The problems are technological ones associated with identifying, locating, measuring, and assembling data and information regarding the locations of land features. They are also associated with people's attitudes and practices in determining how land and its resources are used. Each example includes a solution consistent with American attitudes and practices regarding land, land records, and land governance and examples of opportunities and incentives provided by the full deployment of geospatial technology.

# Notes

1. A land records system encompasses records of both the nature and extent of land features and interests. A land records system is distinguished from a geographic or land information system (GIS/LIS) to the degree that a GIS/LIS emphasizes the spatial location of land features and interests at the expense of the nature of land interests or property rights associated with these locations. The development in recent decades of GIS/LIS has and continues to serve the demand for location-based services. However, these systems are not adequate to serve the demand for records and information about property rights. A land records system is designed to ensure that the records and information about land features are concerned with both the nature and the extent of land features and interests. The design and implementation of a land records system emphasizes the parcel as the common focus for description of rights, interests, and responsibilities associated with land and its resources.

2. A large literature on American land policy and practices exists that includes material on the Public Land Survey System (PLSS). The federal surveyors of the public lands not only measured and marked, they also observed and reported on land and water conditions (Hibbard 1965; White 1983; Johnson 1976). The history of the PLSS is not separable from the history of land law and policy (Clawson 1972; Friedman 1985).

3. The use of the term "land-use drama" is common among lawyers as a description of the public process and venue where land-use planning and management determinations are made (Ellickson and Tarlock 1981, 35).

4. A. D. Tarlock, private communication with Earl Epstein, March 12, 2014. This statement is ascribed to the water and natural resources law professor A. Dan Tarlock. It is a reminder that American citizens always have a legal mechanism for challenging the choice of data, information, and records used to determine their rights, interests, and responsibilities. Many geospatial technology professionals and experts believe that data and information with their imprimatur should control the choice of material used for land planning and management. However, technologies engaging the public are evolving rapidly. Those familiar with the day-to-day land-use drama are aware of the extent to which the choice of data and information used for land- and resource-use determinations are made by representatives, many of them lawyers, of those who have the means to acquire and put forward compelling data and information at the time and place of the drama. The challenge to designers of a modern land records system is discovering how to empower citizens in the process and shift the balance of power among the parties.

# Chapter 2
## Problems created by chaotic land records

Informed land-use planning and management requires appropriate land records and information. The ingredients and attributes of such land records and of the systems that provide this material cannot be separated in substance or process from the demands of all those who participate in the land planning and management process. Provision of the records and information must also be the result of a process perceived to be fair and equitable to all interested parties.

The fundamental dilemma in twenty-first century American land-use planning and management is how to best use land and its resources sustainably and consistent with community attitudes and practices. Aldo Leopold described the problem as, "how to live on a parcel of land without destroying it" (Leopold 1949). Manifestations of the problem range from the global impact of natural and induced climate changes to the local issue of whether a neighbor gets a zoning variance allowing a business in a residence. Population changes, expectations of changes in standards of living, environmentally damaging actions by individuals and groups, inappropriate choices or uses of technology, ethical and religious attitudes about people and the natural world, the choice of economic systems, the choice of political systems, and allocations of the power to use land and its resources assure continuing controversy over land and resource use.

Substantive and procedural demands by those who seek to participate in land planning and management require that records and information ingredients and attributes not be determined solely by the ability to supply land data and information. These ingredients and attributes need to be adapted to the laws and legal processes that define the planning and management process, which reflect preferred attitudes and practices. Actors in the processes have the right to expect the laws and processes to be efficient, effective, and equitable in all aspects.

Geospatial and information technology have altered many expectations regarding land records and information. One altered expectation is that land records and information are reasonably available, at any time and place, and with increasing quality and quantity. This expectation enhances another expectation, realistic or otherwise, that land planning and management will be improved by informed land-use determinations that lead to remediation of land- and resource-use problems.

Land- and resource-use plans, decisions, and actions require knowledge of both the location of land features on or near the earth's surface and the set of rights, restrictions, and responsibilities associated with these features. These rights, restrictions, and responsibilities, called *land interests* or *property rights*, are established by a community and allocated to individuals, groups, organizations, and governments. Recognition and exercise of these land interests are major cultural features of societies.

Informed land planning and management requires use of land records and information about both the nature and extent of land features and interests. However, the ability to connect land features and interests, and the ability to connect records and information with either land features or interests, are far from optimal for satisfying the demands for twenty-first century American land planning and management. The challenge is not only to bring these types of land records and information together but also to ameliorate land planning and management problems in the process. The challenge goes beyond solutions from within the realm of technology. It extends to the realm of arts, social sciences, and humanities where preferred human attitudes are perceived, expressed, and acted on.

## 2.1    Examples of land record problems and their solutions

This chapter presents examples of land record and information problems. Each is placed in the context of land planning and management, and the parties involved require appropriate records and information.

### 2.1.1 LAND TITLES

The term *land titles* refers to a process that yields a description of the location, owner, and substance of land interests (rights, restrictions, and responsibilities—property rights) in a parcel of land. In America, this process involves identification and examination of documents that describe the location and nature of land interests transferred between buyers and sellers, most of whom are private parties.

These documents record the many transfers of land interests that have occurred over many years. Each memorializes an individually negotiated transfer and may contain an independent choice of language to describe both the nature and extent of transferred interests. These documents include deeds, mortgages, easements, and other contracts created in the normal course of a variety of real estate transactions. Copies of these documents are usually, but not always, deposited in the register or registry of deeds, a local government office. These offices are designed to receive, document the reception, index, and archive these documents, and then collect a fee for these services. The register of deeds provides a public repository that gives notice to all in the community that a transfer of interest document has been created and deposited in the office. The office makes no substantive evaluation of the meaning of the document's contents, and the government does not assure the validity of ownership claims made in the documents.

A parcel owner, potential buyer, or other party who wants to know the owner and location of existing land interests initiates what is called a *land title search*. In the United States, private professionals—including title abstractors, private title attorneys, and title insurance employees—do the work. The register of deeds records are searched for relevant documents applicable to the parcel of interest. A private professional examines the documents for their meaning regarding the nature, extent, and allocation of land interests. The professional's judgment and opinion are supported by private assurance, not by the government. This assurance takes the form of a title insurance policy from a title insurance company or the title attorney's private, professional malpractice insurance, verifying that all the appropriate records have been found and interpreted properly. This assurance is valid only for a prescribed term—if the owner refinances a loan, then a new title insurance policy must be obtained. This private, professional opinion is not a determination of the status of land interest ownership equivalent to a judge's opinion after litigation of ownership in a court, an opinion that constitutes a government assurance and generally is sufficient to satisfy American lenders and purchasers. This public-private partnership in the administration and evaluation of negotiated transfers of land interests provides the certainty necessary to sustain the dynamic land market in the United States.

*Specific problem*

A land records and information problem remains. The register of deeds office generally does not record documentation of publicly established land interests—such as zoning ordinances and environmental controls—and government permit actions—such as wetland drainage permits and building code violations. These important public actions affect the nature, extent, and allocation of land interests as significantly as the privately negotiated actions that are described in the documents deposited in the register of deeds.

Documents that describe publicly established land interests are not examined in a typical land titles search process. These documents usually are found in the records of the agencies or legislatures that create them. There is no central repository for these documents analogous to the register of deeds. There is no index or general organization of records of publicly established land interests. If someone wants to know about publicly established land interests, then a difficult search, separate from the titles search, must be undertaken, often with incomplete results that leave a sense of uncertainty.

The public-private partnership that manages the land titles process has problems. The records in the register of deeds do not reflect all private arrangements regarding land interests. Laws do not require all transfer documents to be recorded. Indexes are hard and slow to use because they require alternating searches for the names of buyers and sellers in paper indexes. There is no systematic use of parcel identification numbers (PINs) on documents that relate to a parcel even though modern geospatial technology makes parcel maps and unique parcel numbers easy to generate and attach to parcels.

The American land titles system was developed before the end of the nineteenth century. The basic roles and actions of the various actors in the land titles process—including those of the register of deeds officials, abstractors, lenders, surveyors, buyers, sellers, title attorneys, and title insurance company personnel—have not changed substantially since then. Some of the work within these public and private offices now is done more efficiently because of the application of technology to parcel indexes, archival storage, and private title records and databases. However, the existing land titles system does not connect documentation of privately negotiated allocation of land interests with documentation of publicly established land interests. The widespread use of parcel identifiers on all documents that relate to a parcel does not exist despite modern geospatial technology, resulting in an unnecessarily large document search and management cost and uncertainty about the status of land ownership in all its details. These problems are reflected in real estate closing costs.

At present, it is very difficult to get a reasonably complete and timely answer to this basic question: What can be done or not be done with a parcel of land? This uncertainty remains because the public and private domains of land interest records cannot be connected efficiently, effectively, and equitably.

## Solution

A land records system needs to be developed that identifies and connects records of both privately and publicly arranged land interests. We also need to provide a timely, reasonably complete list of existing publicly and privately established land interests applicable to each land parcel and their allocation among parties for all parcels. This list of applicable documents would not alter the role of the government in the assurance of the documents' meaning and validity. This list would provide means to acquire a complete picture of the nature and extent of land interests at the time and place needed for land planning and management.

To achieve this objective, a registry of documents would need to be developed that records publicly established land interests, analogous to the register of deeds, for privately negotiated interests. This registry's development would encourage public and private institutions to better connect documentation of all land interests, which would provide a better picture of the status of land ownership in the community.

Geospatial technology now makes the creation of this new registry and the connection of its contents with other land records not only possible but also practical. This registry would allow both sets of records to be easily and appropriately examined in determinations concerning who has what power to decide the use of land and resources.[1]

## 2.1.2 BOUNDARIES OF PUBLICLY ESTABLISHED LAND-USE CONTROLS

The boundaries of parcels and areas subject to publicly established land-use controls are not always easily, quickly, or accurately identified in ways that enhance implementation of the intended controls.

Consider an ordinance that reads, "No new buildings within 100 feet of the shoreline." This public action greatly affects privately held land interests and requires location on the ground of the affected area. However, the ordinance does not give legal status to any specific data or map as determinative evidence in the administration of the ordinance. Therefore, a government map or any data or map presented by any party interested in a land-use action related to the shoreline is a preliminary and challengeable representation of the ground conditions. Although it is common practice for officials and citizens to defer to government data or a map, a government map or any other data that has not been explicitly specified by statute, rule, or long-recognized practice as the designated basis for a government action can be challenged, overcome, and dismissed as a basis for agency action. When no specified legal designation of a government map for a specified action exists, an affected party has the legal right to challenge the government map or data at any point in a public land management process.

Challenge to the use of government maps and data in the land planning and management process often happens when the affected party has the resources to gather and present other, allegedly better data and maps. Lawyers capture this concept when they advise clients, "Do not rely on government maps. With good science, boundaries can be negotiated."

*Specific problem*

The absence of specified legal designation or authority for a particular government map or data as the basis for administration of land-use control and similar statutes often generates efforts to locate the shoreline or land feature during the planning and management process. In these ad hoc efforts, evidence is assembled and presented for use at the time of administrative activity. Interested parties can assert that their evidence is better or best. The relative quality of submitted material, including government maps prepared long before the activity under consideration, is determined ad hoc by the administrative official at the time of the administrative action. The wealthy or powerful are able to use the absence of authoritative status for specific data and maps to assert that other material be used to make plans and decisions and take actions. The pressures of wealth, influence, economics, and time affect the choice of data and maps chosen to support the administrative action.

*Solution*

Specific data and maps should be designated as the authoritative material for administration of publicly established land interests. The specification could appear in the statute, administrative rules, or long-established agency procedures. This system would give a priori legal authority to data or maps before they were used in a specific case that implements the public land interest, which would reduce ad hoc conflict over the quality of data and maps. These designations would shift the domain of map and data generation activity away from the time and place of a land-use dispute toward a venue where the agency could make general preparations for all disputes. This system would be a shift from a process often dominated by lawyers toward one dominated by mapmakers and data generators. Although this shift would never be complete, a priori designation of maps and data in this way is a reasonable shift in an appropriate direction.

This concept of designating maps or data for use in specific land planning and management determinations is not new. For example, pilots of large or dangerous vessels must use the specific data and information that appear in nautical charts and supplementary material generated by the National Geodetic Survey of the Department of Commerce's National Oceanic and Atmospheric Administration.[2] No pilot can assert that he or she has other, better knowledge of harbor conditions in any legal proceeding.

Flood hazard maps are also given legally designated authority by local governments for decisions about which buildings are in flood hazard zones for purposes of the federal flood insurance program. Authority is established when a local ordinance, required for local participation in the federal low-cost flood insurance program administered by the Federal Emergency Management Agency (FEMA), is adopted and designates a particular flood hazard map, often the flood hazard map prepared a priori by FEMA, as the basis for the flood insurance program determinations. The designation of authority did not occur when the flood hazard map was made.

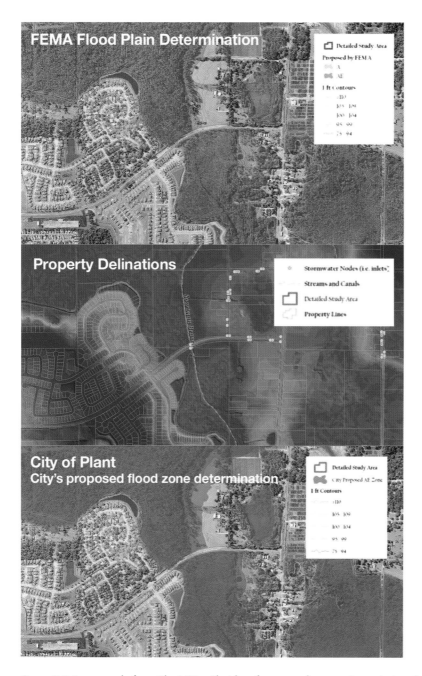

**Figure 2.1** An example from Plant City, Florida, of an appeal process to protest and update flood zone determinations from the Federal Emergency Management Agency's (FEMA) flood insurance rate maps (FIRM).

*Esri Map Book, Volume 24* (Redlands, CA: Esri Press, 2009), 85; courtesy of Brett Gocka and Zlatko Knezevic, Engineering Division, City of Plant, Florida.

Not all land features and publicly established use controls are appropriate for this type of action. Wetlands, for example, may or may not be appropriate for this treatment in some places. This circumstance reminds us of the continued importance of local conditions and determinations of land and resource use. However, there may be some jurisdictions where it is appropriate to designate a particular wetland map or other resource map as the authoritative map for an agency's administration of a local ordinance of the type that states, "No new buildings within 100 feet of a wetland." The basis for this choice may depend more on the need to successfully administer the land planning and management process goals of the statute than the availability of the products of a federal or state wetlands data and mapping program.

## 2.1.3 PUBLIC ACTIONS AFFECTING PROPERTY RIGHTS

Legislatures, agencies, and courts generate statutes, rules, and land-use permits and interpret laws and administrative actions that affect land property rights. Executives issue orders and courts make judgments that also affect land interests. The land records and information that document these activities not only alter land interests but also provide a large untapped source of authoritative land data and information.

*Specific problem*

Consider the example of a hearing examiner for a state's department of natural resources (DNR) who makes a decision in a land-related administrative procedure required by law. The question before the examiner is whether or not a permit should be granted to a riparian landowner who wants to build a dock into a lake. Evidence of the location of the ordinary high-water line (OHWL), the boundary between public and private land, is crucial to the decision. This evidence is presented by officials and other interested parties, and then the examiner makes a decision. In the written decision, the examiner represents evidence of the OHWL by designating an elevation on a topographic or similar map. The written decision with the map depicting the OHWL becomes a public record of the administrative hearing and its outcome. Later, a neighboring landowner on the lake sues the state for creating a cloud on his or her land title. The landowner asserts that the map in the hearing examiner's report so poorly represented the actual conditions on the ground that an interested party reading the report and seeing the map would appropriately conclude that the state incorrectly claimed ownership of most of the plaintiff's land. The issue reaches the state supreme court, and it upholds the neighboring landowner's claim (Zinn v. State 1983).

The problem illustrated by this example is a general one. Maps and data generated during the normal course of agency and court activity regarding use of land and its resources often alter property rights. Regardless of their quality and what they may be considered to represent, some of these maps acquire authority only because they are the result of a public land planning and management action. As in the example, the casual addition of an observation or a combination made with other information to an existing map alters the allocation of property rights. This is a foreseeable problem in many cases.

Geospatial technology has made it easy to make these data and map combinations. Unfortunately, too little attention is given to the impact of these actions and the resulting maps and data on property rights. These data and maps are rarely placed in an open, available central land records repository of public actions, a placement after a formal evaluation by an agency that can both consider the impact on land interests and gain a new source of land records and information. The existence of their data and maps is not active dissemination to interested parties. Landowners have no convenient mechanisms to reasonably identify the public actions that alter their land interests before, during, and sometimes after the action is taken.

Figure 2.2 shows an example of the type of map that appears in connection with an agency's authorized duties. It may be possible to locate the map on the agency's web page if you know that it exists. It is not generally part of a community's registry of public action documents.

*Solution*

A public land records registry needs to be developed that captures land records and information documenting public land planning and management actions analogous to the records of private actions in the registry of deeds. These records and information would be quickly assembled and actively distributed to all interested parties using modern information technology.

This idea is not new. For example, data and information submitted to a zoning or planning commission with the authority to grant a subdivision development permit includes a subdivision plat indicating the intended land divisions. The quality of the plat is subject to scrutiny by the officials. This plat often is sent to the assessor who uses the plat to alter the assessor's parcel map. The plat often appears in documents deposited in the register of deeds.

This action is commonly taken for subdivision plats but not for many other public land planning and management activities that require, use, or create land data. Current geospatial technology removes the technical barriers to assembling these documents in a registry of public actions that includes all actions associated with the subdivision plat. These activities can be used as a source of authoritative land data and information for community land databases.

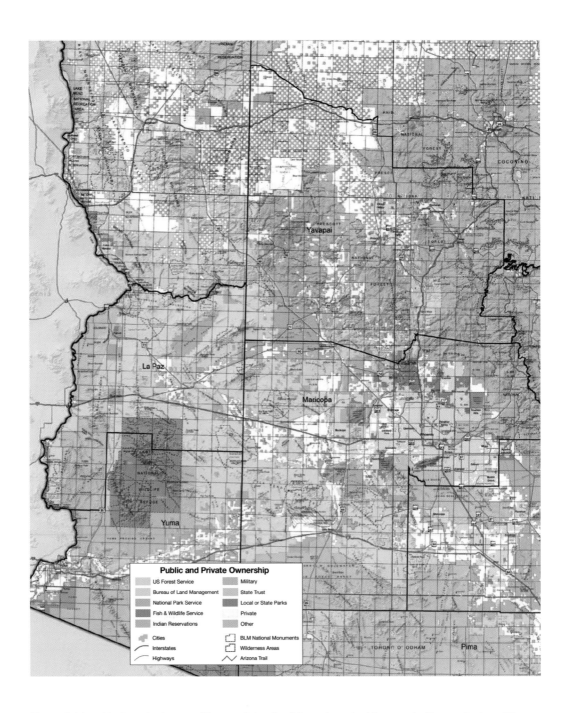

**Figure 2.2** Land in America is owned by a variety of public and semipublic organizations, private entities, and individuals. This diverse and complicated pattern of ownership requires an open and robust land records system. Land shown is in Arizona.

*Esri Map Book, Volume 24* (Redlands, CA: Esri Press, 2009), 54; courtesy of the Arizona State Land Department.

## 2.1.4 MISSING TITLE RECORDS

After the Hurricane Katrina disaster in New Orleans, some homeowners were thwarted in their efforts to acquire rebuilding loans because of land title problems. These were apparent owners who had succeeded parents, grandparents, other relatives, and friends in possession and control of a house and land. Transfer of possession and control occurred when the relative or friend died or otherwise conveyed control without a transfer document that was recorded in the register of deeds, or a title transfer document was created but not deposited in the register of deeds. The current possessor had acted like an owner. The assessor may have sent property tax bills to the parcel address with the name of the previous holder, and the current holder paid the tax bill. The assessor may have changed the name in the tax records to that of the current possessor.

After Hurricane Katrina, the current possessor sought a rebuilding loan from the government or a private lender. This was difficult because the register of deeds records did not reveal that the current possessor held the asserted land interests. Lenders and others rely on long-established practices related to these records. Assessors' records are not considered to be an adequate replacement. Rehabilitation of New Orleans property has been impeded by this situation. This condition of the public ownership records is not atypical in America. Absence of complete ownership data can impede development.

### Specific problem

Assessors' records and maps do not constitute a complete or adequate record of the nature and extent of ownership. They are better described as an approximate record of parcel locations, boundaries, and interest holders that provides the location of the person to whom a tax bill can be sent with the reasonable expectation of a check in return.

### Solution

Existing land interests that are not documented in the register of deeds should be given a degree of designated authority for specified actions. For example, some of the tax assessor's data and information with appropriate attributes should be designated for use after disasters such as Hurricane Katrina. Changes in law and rules would expand the array of records that have a degree of legal authority for use in land planning and management. These records would be identified, given status, and integrated into the land records system.

In the example of the effects of Hurricane Katrina, assessors' records could be specified by statute or administrative rule as appropriate for use in distribution of emergency response funds and benefits. The statute could specify that proper assertion of ownership according to traditional title practices would continue to be the basis for resolution of title disputes.

## 2.1.5 HOUSING RECORDS IN URBAN AREAS

An advocacy group for poor tenants in an urban area seeks to identify and locate the person or persons who own and control the use of poorly maintained, non-owner occupied, and unoccupied housing. At the same time, a building code enforcement officer seeks the owner of a building with code violations in order to deliver legal documents related to the violations. Records in the register of deeds and assessor's offices name a corporation or organization as the owner, with a post office box in another state as the address. The individual with the power to make decisions about the land and buildings cannot be easily identified and located. This problem can occur when city officials consider a law that requires lenders to register vacant, foreclosed homes in city neighborhoods. The law would make property owners liable for a fine and jail if they do not maintain vacant houses and commercial buildings. Both the advocacy group and the agency with the duty to take action regarding vacant and foreclosed homes seek timely contact with the person who has the power to make decisions and take actions regarding the property.

*Specific problem*

Actions by many state and local governments affect the status of land and property in all areas, and these activities are pronounced in urban areas. The records of these actions are available from each agency; however, it is often difficult to assemble a complete and timely record of these actions for a particular property or for a set of properties in an area.

*Solution*

A local, centralized public action land records registry needs to be created to house records and information about both the physical status of parcels and the status of ownership, including documentation containing the name and address of the party responsible for actions at the parcel. This registry would serve as an early-warning system that allows city officials and citizens to track properties as soon as action-related problems are identified.

## 2.1.6 HEALTH DATA

Access to and analysis of spatially specific health data is a complex and controversial issue. Some believe that the ability to locate and identify those with health problems should be carefully limited so that confidentiality is maintained. Publicly available health data generally is reported at a sufficiently small geographic scale so that individuals cannot be identified. Geospatial technology makes it easy to generalize address-specific data so that the detailed location cannot be determined. The underlying data held by public agencies and private organizations may be location specific, whereas the publicly available data are not. Public data are often summarized at the scale of postal codes or census tracts.

Others believe that when citizens suspect that there have been an unusually large number of cases of an illness in their neighborhood, it is too difficult for citizens to acquire data in an effort to examine and verify the suspicion. Public health agencies are prohibited from releasing health data at a geographic scale that could be analyzed to reveal the names of those who have an illness. The concerned citizens have no recourse but to ask the agency to consider the suspicion. They have little or no ability to acquire or analyze the data. The citizens have to rely on the willingness of the agency to execute the analysis, which often is held confidentially because of privacy concerns.

The result of closed and confidential work by agency officials is not always well received. Citizens often feel frustrated when they are not able to obtain data or cannot arrange for an independent data analysis.

Public health officials, private health providers, and concerned citizens would benefit from reasonable access to detailed geospatial health data produced by geospatial software and technology. They benefit from analysis and appropriate representation of the results of the analyses.

Private health providers and public agencies are well positioned to take advantage of modern geospatial technologies for analysis and representation of the distribution of illnesses. For example, a small clinic in Marshfield, Wisconsin, recently asked an expert in geospatial technology at the University of Wisconsin–Stevens Point to prepare maps of the spatial distribution of illnesses that the clinic handled. These maps provide valuable information in a well-received format and aid the clinic in management of several activities. Public presentation of the maps and a description of what they represent could be used to promote the benefits of the studies. However, this is an example of an agency or organization that willingly shares its data with a technical expert in carefully controlled circumstances where the agency seeks help.

Modern technology makes it possible to establish a privately created, publicly available repository of data and records (e.g., a website at which citizens can upload or view health data and records). These data are accessible and available to all. However, the data are limited to volunteered data. The extent to which the data are representative of the general conditions is uncertain.

These health studies—including public agency studies, small organization studies, and public websites—are limited in scope of data and results reporting because the perceived need to assure that individuals are not identified imposes limits.

Although map scales are chosen to secure confidentiality of individual health data, a map such as the one shown in figure 2.3 can generate concern among individuals in the area.

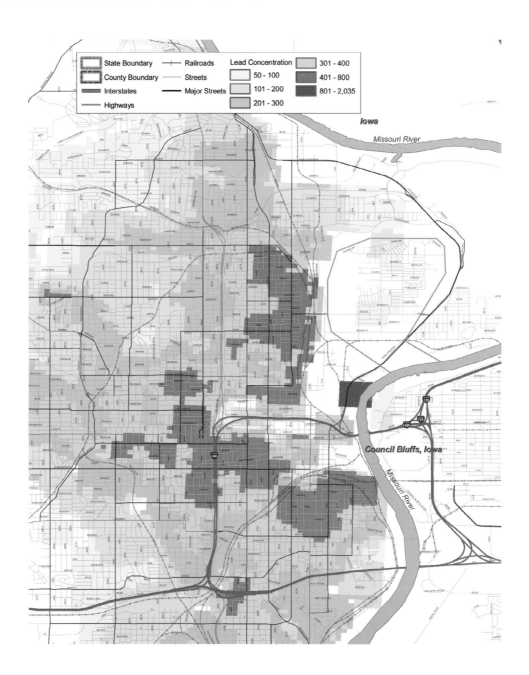

**Figure 2.3** The Douglas County Health Department found elevated levels of lead in the blood of children in east Omaha, Nebraska. Analysis of soil conditions confirmed this finding. These lead concentrations were attributed to a downwind flow from a local lead smelting enterprise. Mitigation involved removing soils with high levels of lead.

*Esri Map Book, Volume 23* (Redlands, CA: Esri Press, 2008), 59; courtesy of the US Environmental Protection Agency.

*Specific problem*

Concerns about protected confidentiality of personal health data require a balance with concerns about local and developing health problems and their sources. Demands for privacy are weighed against concerns about unknown, uncontrolled, and unpublicized activities that contribute to health problems.

*Solution*

The tension between confidentiality and openness of personal- and location-specific health data requires a continuing and balanced reassessment of the existing legally authorized or informal attitudes and practices regarding these data. Existing attitudes and laws provide for tight controls on access to health data held by both public and private health organizations and individuals. Open records and freedom of information laws in most jurisdictions have provisions exempting personal health data held by public agencies from openness. Laws governing the professional actions of private health providers afford similar protections.

It is possible for health researchers to obtain data for health studies from public and private sources. This access is not available to the general public, who often has no knowledge of its existence. The identities and qualifications of those who have access generally are not known.

The issue of access to personal health data remains a difficult one. It is easy to generalize address-specific data so that the actual location on the ground cannot be identified. However, it is possible to expand the domain of those who have controlled access in a carefully prescribed way by statute, rule, and professional regulation. This expansion should be a subject of public debate to educate the public on the advantages of better health studies and to promote the general sense that government agency health studies are reliable because external opportunities to examine government actions and studies are readily available.

## 2.1.7 HISTORICAL DATA

Many land planning and management activities are aided by records and information concerning the historical condition of land and its resources. For example, historical data concerning the nature and extent of wetlands can inform watershed planning and management. Similarly, historical data concerning the types of trees at the time of agricultural settlement can inform land planning and management when there are observations of an allegedly remnant species.

*Specific problem*

Consider the example of citizens and officials involved with the planning and management of land and resources at or near a lake. The status of features—such as plants, wildlife, trees, and soils—are a concern. Interested parties observe a stand of trees no longer common to the area but suspected to have been common at the time of agricultural settlement. These parties want to identify, locate, and use any historical data that support the assertion that the currently existing stand of trees is a remnant old-growth stand. However, current pressures to make a timely development decision about land use may preclude the time and costs of a search for historical data. The development decision will then be made with existing, easily available data.

Historical data are often in the form of paper records and maps stored in musty archives, which discourages their use. Geospatial technology development has encouraged the creation and use of digitally stored material. Increased use of digitally stored historical data involves the operational costs of identification, location, assessment, acquisition, and conversion, which can appear daunting.

This organizational issue can be addressed by asking the following question: Are there identifiable issues in a community that encourage the investment of time, effort, and resources in identifying, locating, and converting into digital form those historical records and maps that may be a valuable resource in future land planning and management? Although the answer to the general question is specific to each community's local conditions, in many cases, the answer is yes.

A major example of this potentially valuable material is the set of records and information created as the Public Land Survey System (PLSS) was established across much of America. The surveyors of the public lands measured, marked, and described their spatial measurements. The surveyors also observed and recorded information about the characteristics of the land and resources in the vicinity of demarcated points.

The ability to identify, locate, acquire, and use these PLSS historical records varies greatly among states and jurisdictions. Wisconsin maintains easily accessible digitized PLSS records that include both survey measurements and observation of land and resource conditions at the time of the surveys (Mladenoff 2009). In Ohio, the original paper PLSS records have not been digitized and have been moved to the state archives. Efforts to identify, locate, copy, digitize, and distribute these records for one or a few PLSS points are time consuming and expensive. Their use in current land planning and management is significantly deterred.

PLSS records and information are just one set of important historical records that are useful for current land planning and management. Title and land interest transfer documents are already centralized and organized in register of deeds offices. However, other historical records that are scattered, disorganized, and effectively unavailable are or may be predicted to be important material for future land planning and management. This information includes historical records of wells and their locations, wetlands, cemeteries, abandoned or converted industrial sites and activities, and riverbeds.

*Solution*

Modern geospatial and information technology should be used to efficiently, effectively, and equitably identify, locate, digitize, and distribute historical records, relying on identified, predictable community needs for such material. Records of long-established water wells in areas where hydrological fracturing (*hydrofracking*) occurs or will occur are an example of such needed material. The extent of need for these historical records is indicated by the potential for hydrofracking, as shown in figure 2.4.

## 2.1.8 INVASIVE SPECIES

Invasive species become a social and community problem when they are seen and perceived to be undesirable. These observations and perceptions may or may not require immediate, individual, or collective remedial action.

These observations, attitudes, and practices become a land records and information issue when citizens, groups, organizations, and officials ask the following questions:

- What do the observations represent?

- Are they an invasive species?

- Have they been properly identified?

- Do the observations represent a threat to the community?

- Should remedial action be left to individuals?

- What remedial actions are appropriate?

Citizen observations can now be deposited and distributed easily to the public. A photograph of an allegedly invasive species made with a mobile device and with an attached set of geospatial coordinates can be uploaded to a variety of widely disseminated and accessible databases and websites. This "crowdsourcing" is happening in many areas. The data does not need to be authoritative to be valuable. Existing technology can verify that the data came from a specific citizen, with the data time stamped and location specified (to prevent location spoofing).

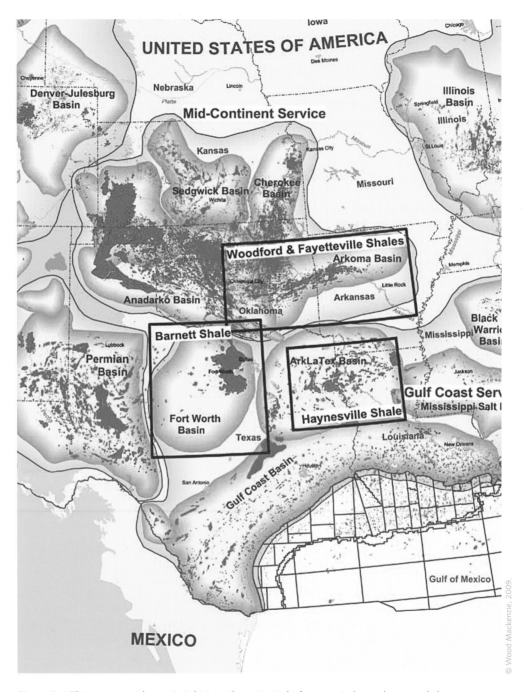

**Figure 2.4** This map provides an insight into the potential of energy independence and shows a large-scale view of three newly identified sources for oil and gas producing areas in the midcontinent area of the United States.

*Esri Map Book, Volume 25* (Redlands, CA: Esri Press, 2010), 90–91; copyright Wood Mackenzie, 2009.

Citizen observations, such as those for invasive species, require professional, expert verification of their authenticity and impact before collective remedial action is undertaken. Individual action by a landowner risks negative impacts on a neighbor's land.

### Specific problem

Community remedial action through agency activity affecting many properties depends on expert verification and evidence of widespread distribution of the invasive species. Community remediation actions (e.g., spraying by a government agency) require a process wherein the agency acquires a set of data, assesses the quality of these data, presents the results to the community, and proposes appropriate action. This process is part of the typical land management procedures in a community.

### Solution

Citizens need a formalized, administrative process where they can deposit observations, such as those of invasive species, and have these observations actively received and considered for authoritative designation for agency action. These observations could then become a part of a community land planning and management program designed to ameliorate the problem. The emphasis is on the citizens and the officials in a joint effort to collect and assess the data that will be used by the community to address the problem.

In addition to a place where the observations can be deposited, an administrative process should be established that both receives and gives authenticity to appropriate observations to promote efficient, effective, and equitable administration of remedial activity.

Those responsible for this activity should be able to receive, evaluate, and distribute the material to the community, gain the community's support for remedial action, and ensure professional assessment of all the data, its representation (e.g., as maps), and presentation of the analysis and supporting maps to the public as basis for community remedial action.

Representation of the all data, including data generated by citizens and considered by officials, improves the prospect of community support for remedial actions, such as spraying, that affect many landowners and community members.

## 2.1.9 SCENIC AND CONSERVATION EASEMENTS

An easement is an interest one party has in the land of another. Commonly, easements are rights-of-way that allow a party (e.g., an electric utility) to have the right of access and use some part of another's land.

A scenic or conservation easement is a relatively new land interest. It is a land interest held by a party (the trustee) that limits a landowner's actions to those that do not interfere with visual or other natural aspects of the land. The trustee can be an individual, a public or private group, or a government agency.

A scenic or conservation easement gives the trustee a responsibility to assure that the landowner and successor landowners do not take land-use actions that interfere with the goals of the easement agreement. The trustee is interested in observations and other data that promote the trustee's ability to sustain the easement's conditions over time and thereby fulfill the trustee's responsibilities. Maintenance of the trust conditions requires timely notification to the trustee and other interested parties of observed and recorded changes in both land ownership and activities that violate the trust conditions.

### Specific problem

A landowner may knowingly take actions inconsistent with the easement terms. Trees may be cut down, resulting in a significant change in the landscape. Alternatively, the landowner may not be fully aware of or understand the limits imposed by the terms of the easement, even when the landowner knows of the easement's existence. Landowner actions inconsistent with the conditions of the easement trigger a response by a trustee responsible for compliance with the conditions.

The trustee needs a variety of records and information about the status of land conditions and the terms of the easement. Observations can be by the trustee if the trustee is regularly on or near the land. Often, the trustee is not regularly near the land and relies on other data and observations. These observations must be conveniently and accurately connected to the parcel to which they apply, connected to the fact that there is an applicable easement, and available to all interested parties, including the public. Failure to monitor ground conditions, including changes in ownership, in a timely and complete manner establishes conditions that make it easy to violate the trust objectives.

Consider the costs of untimely notice to the trustee of changes in land ownership. The National Park Service (NPS) in the US Department of Interior administers the National Wild and Scenic Rivers System.[3] This program includes the St. Croix River in Wisconsin and Minnesota. The NPS is trustee for many scenic easements along the river in two states and many counties. A highly trained NPS specialist in land planning and management for this type of natural resource is required to spend a significant time each year going to register of deeds offices in each of the many counties along the river in order to update changes in ownership of those parcels subject to scenic easements. This not only consumes a lot of time that could be used for professional management efforts, it also delays attention to NPS efforts to sustain the scenic easement provisions.

Private land trusts that hold scenic and other environmental easements face similar challenges and costs with efficient and effective execution of their duties.

*Solution*

Modern geospatial and information technology make it easy for the register of deeds to actively and immediately notify interested parties of recorded easements. Digitization of the register of deeds indexes and recorded documents is required for modernization of these offices. These indexes need to be parcel based or have some means to connect the document to the land. This connection makes it efficient, effective, and equitable for the register to notify all interested parties electronically and in a timely and complete manner of document recordation and its type. Efficient, effective, and equitable land planning and management is enhanced by prompt action based on timely notice. The NPS official can then be more productive.

Records about the land interests described in the easement and ground conditions include a notice to the trustee of a parcel ownership change. The trustee can inform the new landowner in a timely manner of the trustee's understanding of the easement's conditions and of the trustee's intended actions regarding sustenance of the trust duties. Prompt notice to the new landowner of the trustee's understanding, obligations, and intentions promotes compliance. Prompt notice to the landowner requires prompt notice to the trustee. Prompt notice to the trustee of a change in parcel ownership is not the usual circumstance unless the trustee is frequently near the parcel. Instead, most transfers of ownership occur without the trustee's knowledge until long after the transfer. This failure to notify trustees creates problems in maintenance of the trust conditions.

The benefits of timely notice often extend to the community and others outside the register of deeds office. This extension of benefits requires that both the register of deeds and the community think broadly about the community-wide impact of register of deeds activities, not just from a perspective within the traditional register of deeds. If the register of deeds and the community restrict their attention to traditional actions within the office, and thereby think within the institution, many of the benefits of modernization will not be captured.

Prompt, active, and widespread dissemination of property transfer documents recorded in the register of deeds is not limited to scenic easement documents. Dissemination of all its records in this way can garner similar benefits.

In the twenty-first century age of geospatial and information technology, register of deeds offices need not retain their nineteenth-century practice of passive response to record requests. The office can look outward, devote attention to community needs, and generate significant benefits for their communities.

## 2.1.10 PRIVATELY RECORDED MORTGAGE RECORDS

Gaps appear in the set of register of deeds' title records for more than one reason. One is the failure of owners, for reasons described earlier, to record a title transfer document. Following are other reasons for gaps.

*Specific problem*

The recent development of the practice of not recording subsequent sales of mortgages in the secondary mortgage market and the creation of bundled mortgage financial instruments based on these mortgages lead to gaps in title records.

The Mortgage Electronic Registration Systems (MERS), a private company, has become the mortgage company of record for mortgage documentation in the register of deeds in the jurisdiction where the land parcel is located.[4] Records of subsequent sales of the mortgage are handled by MERS, not by the register of deeds. The name of the mortgage holder, who has the power to foreclose, is not easily available to the landowner, local community, or securities rating services. The asset is separated from its record, resulting in hidden information that acquires its own economic value separate from that of the asset. The issue is how much of this hidden information is appropriate for a community.

Mortgage lenders suggest that this private record keeping is more efficient than public recordation in the community where the land is located, especially for subsequent sales of the mortgage. It was argued that subsequent mortgage sales often involve a sequence of buyers and sellers far removed from the public registry. This disconnect, it was argued, made the MERS operation an efficient alternative.

The MERS operation is reasonable as long as there are traditional rates of mortgage terminations by the foreclosure process. A nonforeclosure termination of a mortgage involves a willing buyer and seller. In this typical situation, the status of intermediate mortgage transactions is not a concern to the willing buyer and seller. The seller is concerned that the mortgage be satisfied and that a document describing this final action is deposited in the local registry.

However, mortgage foreclosure involves an unwilling owner protected in the foreclosure process by public laws. When the rate of foreclosures exploded during the foreclosure crisis, the MERS operation could not cope with the rate of detailed documentation required by the foreclosure protection laws. These laws provide for careful documentation of each step in a mortgage's history. In order to meet the rate of mortgage foreclosures desired by the lenders, many foreclosure documentations were done in a manner that did not meet foreclosure law standards. This formed the so-called robo-signing that eventually exacerbated the foreclosure problems.

The recordation of subsequent mortgage transactions outside the public records system has efficiency aspects for some lenders; however, document recording fees fund public registries. Development of MERS greatly affected registry revenue, which was needed for modernization. The public aspects of foreclosures, represented by foreclosure laws, and open public records in the local community where the land is located raise issues of effectiveness and equity in land planning and management.

*Solution*

The general practice that all real estate transactions, including subsequent sales of mortgages, be recorded in the title records repository in the community where the land is located should be retained. Retention of this traditional practice also sustains the revenues in the form of document recordation fees for the local public records office (usually the register of deeds) that are now lost to the private MERS activity. This practice also promotes the traditional public notice procedure associated with the register of deeds.

Modern geospatial and information technology make this retention of practice practical. It is possible to distribute records of mortgage and other land interest transfers to local registers of deeds from anywhere in the world.

## 2.1.11 LAND RECORDS AND PARCEL MAPS

Maps of all parcels in a community are most often created and sustained by the property assessor's office in American local jurisdictions. Officials rely on parcel descriptions that appear in documents, such as deeds and mortgages, deposited in the register of deeds during the course of property transfers. The descriptions summarize observations, measurements, and demarcation of the location of parcel boundaries. Many of these descriptions are the result of actions by an agent, usually a professional surveyor, for one of the parties to the transaction. Sometimes, a lawyer for one of the parties writes the description (figure 2.5).

Geospatial technology makes it easy for the assessor's office to acquire and examine the description in a transfer document. This official, perhaps aided by the work of a professional in the engineer's office who has knowledge of and experience with surveying standards, makes a judgment of the description. A satisfactory judgment leads to incorporation of measurements in the assessor's local government parcel map. This technology also makes it convenient to compare the map representation of the new parcel description with the map representation of the parcel in the existing parcel map. The two representations may not be the same. This technology allows the official to adjust any discrepancies and produce a clean-looking parcel map.

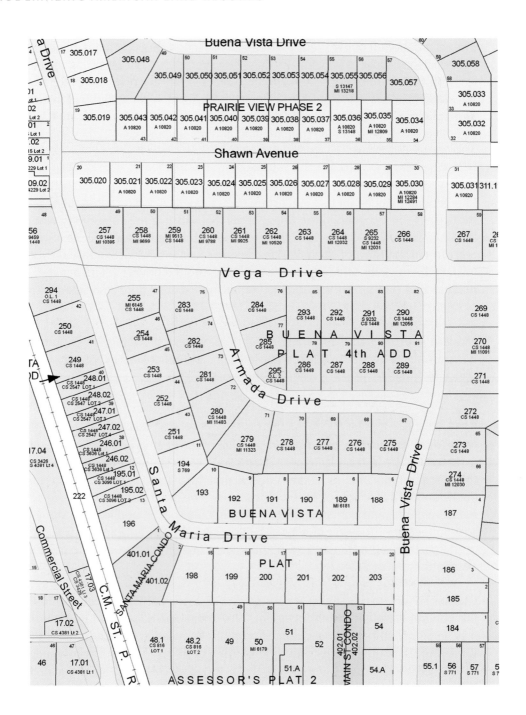

**Figure 2.5** Columbia County, as part of the Wisconsin Land Information Program, has helped set standards for the annual production of tax parcel maps as part of its tax roll workflow process in accordance with state law.

*Esri Map Book, Volume 25* (Redlands, CA: Esri Press, 2010), 65; courtesy of Columbia County Land Information.

It must be noted that the official who assembles the data or makes the map does not have power to give legally recognized authority to the observations, measurements, or maps as determinative of parcel boundaries. Instead, the resulting local government parcel map is a representation of the approximate parcel boundaries. This map is, in many American jurisdictions, adequate for many location-based services, such as vehicle routing, navigation services, spatial representation of street addresses, and representation of other land features. This map is not authoritative for a determination of parcel and land interest boundaries that is legally binding on all parties associated with the boundary. The local government parcel map is not appropriate for resolution of a boundary dispute between neighbors.

### Specific problem

The issues regarding local government parcel maps prepared in the manner described before and in common practice in America are the following:

- Are there local, regional, or national reasons for increasing the authority of these maps in the process of establishing legally binding parcel boundaries?

- If the answer is yes, then how can this be done given the prevailing attitudes and practices?

We should never forget that maps are a representation, however good and accurate. Boundaries are represented on a map but only occur on the ground and are determined by best evidence of their location on the ground. American surveying law and practice give precedence to best evidence of the location of boundary features, such as monuments and fences, in their original location (Brown et al. 1986). Precedence is not given to locations represented on a map.

### Solution

The need for a representation of all parcels and their boundaries and the small number of seriously contested boundary disputes requires that all reasonable efforts be made to establish complete, local government parcel maps in most if not all jurisdictions. These maps would adequately serve most of the demands for location-based services. Geospatial technology development would allow digital versions of these maps to become the standard form. This local government parcel map should be the basis for an index to all land records that relate to the parcel, including records that describe the nature of land interests in and the location of parcels.

There may be communities or areas where there are reasons to establish a local government parcel map that contains data and information used in legally binding resolution of boundary disputes. Some boundary disputes between neighboring, disputing parties are taken to courts for resolution. The court results, including the boundary descriptions, are binding on the parties. The court-determined boundary description has been given a legal authority for boundary purposes. This description can be forwarded to the assessor or local government mapmaker who can flag this description and assure that it is given priority over other descriptions whenever a new map is prepared.

The domain of government actions that generates authoritative boundary data and maps can be expanded beyond the activity of the courts. For example, standards and practices in agencies, such as an engineer's office, empower them to act as dispute resolution agents to whom disputing neighbors can bring their dispute and their boundary evidence. The results of this process can be given authority and designated for priority use when a new parcel map is generated. The results and designation may not give them authority for final resolution of boundary disputes; however, an increased file of designated records and information for purposes of a local government parcel map is developed and expanded.

## 2.2    The nature and scope of the problems

The previous examples illustrate how existing, independent land-records-information institutions inadequately serve the needs of many who seek to participate in a community's land planning and management. This assertion is not based on how well location-based services are supplied by the products of geospatial technology. Instead, it is based on the incomplete deployment of that technology to satisfy broader demands for a more complete, organized set of records and information. This inadequacy imposes significant costs on all land-related activities, including but certainly not limited to real estate closing costs.

Many citizens, groups, organizations, and officials are thwarted in their efforts to assemble and use land records and information in ways that promote their full participation in land governance.

The independent institutions have long-standing patterns and practices that make it difficult for them to cooperate with one another in the assembly of material needed to address land- and resource-use issues in the community.

To summarize, the problems with existing land records and information systems are the following:

- Records that document observations of the location of land features are not well connected to records of the nature of land interests associated with the features.

- Records of privately created land interests are separated from records of publicly created land interests.

- Long-established attitudes and practices in both the public agencies and private organizations that manage these records create a set of largely independent land records institutions.

- Records and information that inform interested parties about both the nature and extent of interests in a parcel are difficult to identify, locate, acquire, assemble, and apply in the process of land planning and management.

- It is difficult for citizens and groups to deliver land records and information to officials in agencies and organizations who not only receive the material but also have established, recognized processes that consider the material and give it authority in the normal course of agency duties.

- Private land records institutions have symbiotic relationships with public institutions, which add to the complexity of designing changes to the land records systems.

Citizens of some American communities favor their increased participation in land planning and management. These communities are most likely to use the most modern geospatial and information technologies to increase access to materials in the several public land record institutions. These communities are also most likely to succeed in overcoming the institutional barriers to connecting the different types of land records. The demand for land records and information by all actors in the many encompassing venues for land planning and management constitute a force for change.

## 2.3   A caveat about long-established land records institutions

Each independent land records and information institution (register of deeds, assessor, title insurance company, etc.) has or seeks land records and information with attributes appropriate for satisfaction of its mandated, authorized, and market-driven functions. For public institutions, these data attributes are determined primarily by the enabling legislation, administrative rules, court decisions, professional standards, and accepted practice among similar agencies.

The records and information within the private-sector institutions have attributes determined primarily by the needs of targeted customers. Private institutions often gather data from public and other private sources and make assessments, changes, or additions in order to satisfy their customers' needs. A title insurance company is a good example of this private activity. This symbiotic, public-private partnership is often necessary because the public institutions do not provide the data connections and analysis that private companies provide to their customers.

Long-standing, public-private partnerships are a significant aspect of existing institutions. Connecting records and information is often a major function of the private companies. Modernization of land records systems cannot be expected to overturn this partnership. However, the demand for appropriate material in twenty-first century American land planning and management can be an incentive for change that serves both public and private activities. The symbiotic partnership can be both an incentive and a barrier to changes in the nature and extent of the overriding land records institution.

## 2.4   Challenge

The fundamental institutional problems, illustrated by the examples presented in this chapter, create a substantial challenge: overcoming the barriers that inhibit appropriately connecting two different types of land records. One type is data and information about the location and extent of land features on or near the earth's surface. The other type is data and information about the recognized rights, restrictions, and responsibilities that adhere to these features.

Adding to the challenge is that the two types of land records differ in a fundamental way. The location and extent of land features can be observed, measured, and represented. However, land rights, restrictions, and responsibilities are not equally observable and measurable. The former can be seen, whereas the latter are unseen but powerfully perceived.

The difference between things seen and unseen has long been a matter of human interest. Long ago, it was written that "Things which are seen are temporary, but the things which are not seen are eternal" (2 Cor. 4:18, New King James Version).

Similarly, the emotionally powerful and extremely popular World War II song "I'll Be Seeing You" captures differences between things seen and things unseen. A first part directs attention to things seen, such as a café, park, carousel, and tree. The lyrics then turn to the human ability to see someone in a lovely summer's day, in things light and gay, and in a vision of the moon.[5]

The difference between what is seen and not seen raises questions about what is more powerful—images or what people "see" in images—and how long-lasting and powerful is the difference. Is a picture worth a thousand words, as the saying goes, if a particular image does not capture the effect of the words? What specific picture captures the sense of what is observed when a person looks at the moon and responds in the way that the narrator does in "I'll Be Seeing You"?

The emotional impact of this song comes from connections between what is seen and what is unseen. Similarly, land ownership in all its aspects is, by its nature, an abstract concept. It cannot all be seen. Although the basis for rights, restrictions, and responsibilities can be described, their real meaning requires more representation. Modernization of land records and information comes from better connections between records of what can be seen and records that represent and designate what cannot be seen.

The hierarchy of data, information, and knowledge reflects the nature of the difference between things seen and unseen. Data such as the location of land features are considered observations. Information is the result of placing these observations in a specific context created by a map with its distinctive scale and symbols. Knowledge is created when an individual views a map and sees the connection between what appears in the map and the individual's unique understanding and sense of that image.

Maps of the boundaries of land interests, called *parcel maps*, are necessary. These maps are the bases for connections between the parcels and all the documents that describe the nature and extent of land features and interests. The unique PINs on these maps make these connections possible.

## 2.5   Conclusion

Citizens, groups, organizations, and officials face many problems in their efforts to identify, locate, acquire, combine, and use the full array of land records and information in the normal course of their participation in twenty-first century land planning and management.

A properly constructed land records and information institution would balance individual, group, market, and state interests in land. This institution would provide for informed citizens with connections to others. This balanced set of interests, and the land records and information system that supports it, would form a larger, collective understanding of rights, restrictions, and responsibilities in a modern society. When data management is seen in this way, the land records and information institution in a community will become a means for community engagement in land planning and management.

It can be argued, for example, that the inability to attach mortgages to the community parcel maps inhibited the ability to see the escalating nature of the real estate crisis.

Geospatial and information technology successfully provides many location-based services. The full deployment of these technologies in the community is impeded by institutional barriers preventing interested parties from connecting land records and information about both the nature and extent of land features and land interests.

A two-way land information portal allows citizens to both submit and withdraw records and information to and from agencies. This portal makes it possible to connect citizens, neighbors, and the community to aspects of land that matter to them, promoting citizenship and empowerment. This portal goes beyond parcels, maps, and land interests to citizen governance. Seen in this way, development of a modern land records institution is a model for development of modern American society.

Before details of a solution can be explored, it is appropriate to examine the origins of the arrested state of developed American land records and information systems. This historical analysis is necessary because it reveals much about American attitudes and practices that establish both incentives for and barriers to change. Chapters 3, 4, and 5 consider these attitudes and practices in the related histories of land, land records, and land governance in America.

## Notes

1.  The establishment of land-use controls and other publicly created land interests is a feature of American land planning and management. This process began in earnest at the beginning of the twentieth century and is most pronounced at the local level of government. These interests—sometimes called *overriding interests* because they influence and often override privately established interests—are routinely a part of the record of interests in a cadastre found in other countries. The need for a registry of these publicly established interests was clearly identified in the recommendations in *National Land Parcel Data: A Vision for the Future* (National Research Council 2007).

2.  The system and legal process for production and use of nautical charts has a long-established feature that gives overriding and incontrovertible status and authority to the data and information associated with the charts. Pilots of large vessels or vessels with dangerous cargoes are required by law to make piloting decisions based exclusively on the recognized data band information included in the charts. No other data and information can be used, *regardless of how much the pilot knows that material is not correct*. For example, a chart that shows an area within which

a vessel sank in 1800 and over which many ships have passed without incident is an area forbidden to the pilot as long as the area continues to appear on the chart. No data, information, and knowledge external to the chart system are admissible as evidence in a court proceeding where piloting activities are scrutinized. The chart material is authoritative for pilots. This type of legal regime for data, information, and knowledge is a basis for design of a land records system that seeks to identify material for a complete record of land features and interests. For an introduction to the American system of nautical charts, see the National Oceanic and Atmospheric Administration (NOAA) Office of Coast Survey Nautical Charts and Publications web page at http://nauticalcharts.noaa.gov/staff/chartspubs.html.

3. Successful implementation of planning and management objectives in a program, such as the federal National Wild and Scenic Rivers System, depends on timely and appropriate data and information at the parcel level. The federal government has acquired a scenic easement and a duty as trustee to enforce the terms of that easement for a large set of parcels strung out in a narrow band along the river. The federal officials need timely data and records from local government offices, especially the county-level register of deeds. These officials need to know, among other things, when parcel ownership changes. Trust duties are enhanced when the federal official can inform the new owner of both the owner's and the trustee's rights, interests, and responsibilities. Timely communication promotes implementation of the easement's objectives, a benefit based on the observation that "an ounce of prevention is worth a pound of cure." Currently, federal officials professionally trained to administer the complex physical aspects of a wild and scenic river must spend extraordinary amounts of time and energy (literally and figuratively) periodically going from registry to registry in a search for parcel ownership changes. This activity remains necessary in the twenty-first century because the registry is a passive office with nineteenth-century processes. The registry's records are open and available to all who come to its premises and examine the records there. In a twenty-first-century activity, the registry could electronically deliver notice of the recordation of a warranty or similar deed to the National Wild and Scenic Rivers System's management office. The benefits are in the form of enhanced program implementation and reduced costs to the program in the form of time and energy expended to maintain records of ownership. The problem is that these benefits accrue to the community, not directly to the register of deeds. For an introduction to the National Wild and Scenic Rivers System, see its website, http://www.rivers.gov/.

4. The privately operated Mortgage Electronic Registration System (MERS) has been a significant object of public attention since the collapse of the so-called real estate bubble and the subsequent foreclose crisis. The activities associated with MERS are especially important to the subsequent sales of already existing mortgages and, especially, the sales of bundled mortgages where the sales are far removed from the locations of the parcels. The creation of MERS in the mortgage lending business represents a removal from the public title records to the domain of private records of the documentation of subsequent sales of existing mortgages. The foreclosure crisis was enhanced when the MERS operation had to deal with the requirement in many states that foreclosure proceedings document subsequent sales of mortgages with documents from the public offices. The development of MERS can be viewed as a step in the process of removing some, or potentially all, privately arranged title transfer documents from the domain of public title records, thereby obviating the need for local register of deeds offices. MERS and the potential of MERS is a major change in traditional practices worthy of significant discussion (MERS 2013).

5. Irving Kahal and Sammy Fain composed "I'll Be Seeing You" for the "Royal Palm Revue." We have relied on the performance of the song by Tommy Dorsey and Frank Sinatra, recorded on February 26, 1940, reissued on the album "Stardust" by BMG/Victor/Bluebird, 1992. Bluebird 07863. 61073-2.

# Part II

## History of land, land records, and land governance in America

Part II examines three related histories that are important to understanding how and why American citizens use land records and information in the planning and management of land and its resources in twenty-first-century America: 1. land in America, 2. American land records and information institutions, and 3. American land governance.

Chapter 3 focuses on the history of the concept of land in America. This history suggests that although many would like to say, "It's my land and I can do what I want with it," land in America can be described as a commodity affected by a public interest.

Chapter 4 focuses on the related history of land records institutions. The important observation is made that Americans have emphasized a public/private partnership for collection, storage, organization, and analysis of privately arranged transfers of land interests while neglecting establishment of a system for important publicly established land interests.

Chapter 5 considers the history of governance, particularly land governance, in America. This history reveals that from the time of its early English settlers, Americans have sought to participate as directly as possible in the governance of their lives. Notwithstanding the stresses of the modern world, American citizens continue to seek this direct, participatory governance, especially concerning determination of how land and its resources are used.

These chapters encourage examination of long-established, preferred American attitudes and practices about the use of land and its resources and the processes for determining these uses. The related histories indicate how these preferred attitudes and practices have influenced the development of existing American land records and information institutions. They also reveal attitudes and practices that are both incentives for and barriers to modernization of the land records and information institutions.

These histories inform efforts to modernize land records systems and fully deploy geospatial technology by placing these efforts in the larger context of twenty-first century American land planning and management. This context offers a view of the large demand for records and information and the supply of new products, a demand that can be used to encourage, direct, and sustain the modernization.

# Chapter 3
## Land in America

The concept of land in America includes attitudes and practices about how land and its resources should be used. Recognized patterns of preferred thought and action appear as land law and legal process that influence all public and private land planning and management.

History provides a means to identify and describe the mainstream attitudes and practices regarding land and resource use, procedures, who has the power to make determinations, and who benefits. History also gives an important indication of the strength of the preferences and the likelihood of the community to undertake changes in long-established thoughts and actions.

Each society has its own, sometimes unique, land concept. Even within a society, people think of land in different ways. A suburban homeowner may focus on the highway and road configuration that determines the commute to and from work. An owner in a remote, recreation-oriented area may be interested in preservation of community resources and scenic amenities. An inner-city resident may be concerned about public and private commitments to nearby residences that determine safety in the area. Real estate developers are interested in the potential value of parcels if and when they can be devoted to new uses and in the community's willingness to support these new uses. Sometimes the diverse set of attitudes and practices coexist in an American society that gives great deference to private determination of land and resource use.

Questions such as the following arise:

- What is the American land concept in the twenty-first century?

- What are the origins and nature of the American land concept?

- Have attitudes and practices changed?

- What are the direction, extent, and nature of change?

- What processes are preferred for land-use determinations?

- Who determines land and resource use?

- What land records are appropriate for and consistent with the American land concept?

The definition of land includes the following:

- The solid part of the earth's surface not covered by water.

- A specific part of the earth's surface.

- A country or region, such as a distant land or one's native land.

- Ground or soil in terms of its quality and location, such as rich land and high land.

- Ground considered as property, an estate, or a specific land holding.

All but the last item concern the location of observable land features. Ground as property concerns aspects of land that are not easily observed and measured. These aspects are land interests—the allocated rights, restrictions, and responsibilities that establish who has the legally recognized power to determine use of land and resources in a community.

Controversies arise when individuals, groups, officials, and entrepreneurs have different perspectives on not only how land and resources are used but also on how use determinations are made. Laws and processes formalize the preferred patterns of land administration and governance in a community.

Behind every land conflict are two fundamental questions: What is the appropriate status of a piece of land? Who has the right to decide (Andrews 1979)?

Answers to these questions are found in observations of the location and extent of land features on or near the earth's surface and in the community's attitudes and practices regarding the features. The preferred attitudes and practices often are long established in a community and are part of its history.

# 3.1   Precolonial and colonial land concepts

The division of land into parcels began as societies evolved from hunting and gathering to farming. Ancient Babylonian texts describe parcel boundaries and land interests (Brown et al. 1986). Creating parcels allocated to some, and not others, the power to determine use of the parcels and their resources. This allocation became a means for consolidating power in a community. For millennia, the allocated ability to determine land use meant power (Worster 1985). This power remains important in the modern world.

### 3.1.1 ENGLAND

The concept of land in America has origins in Europe's change from the feudal system of allocating land interests that was established after the fall of the Roman Empire. The system of English land rights, restrictions, and responsibilities that followed the Norman conquest of England in 1066 AD is considered feudal (Hallam 1986). In that feudal system, a person received land interests or property rights in subordination to a superior person. The junior person owed inescapable duties to the superior in exchange for the rights allocated to him by the superior person, ultimately the king. Social, economic, and scientific forces already at work at the time led to changes in this system of feudal land interests in both England and elsewhere in Europe. In England, the process began early, with a powerful and sustained momentum. In 1215 AD, the Magna Carta changed many land-based relationships between the king and English barons and other recipients of feudal land interests. The first substantive clause in that document dealt with land interests. This clause allowed barons to transfer their land rights to their chosen heir without approval of the king. Although this authority appears unremarkable to us today, prior to this clause, land rights and the authority to choose a baron's successors returned to the king after a baron's death. This clause represented a new relationship between the king and the barons.

Over the next several centuries, more rights to determine land use were wrested from the king by the barons and others. Increasing numbers of the English population gained these rights, free of feudal duties or subordination. By the seventeenth century, it was possible to refer to the English system of land interests as something other than feudal.

A system emerged wherein many if not most powers to determine land and resource use without royal approval belonged to private individuals or groups. Only a few powers remained with the king (the *Crown*, or *government*, in modern parlance). The important powers were the property tax, the taking of power (now called *eminent domain*), and the ability to control land use (now called *land-use controls*).

**Figure 3.1** Traditional private and public land interests.

The result of this historical development is the system of land interests, or land tenure, that prevails in modern America, called the *freehold system of estates*. This system is represented in figure 3.1.

The specific land interests are represented in the figure graphically as a bundle of sticks. Each stick is a legally recognized power to determine a specific land or resource use, without government approval. A stick may be allocated to a private party or to the government. It is traditional to think of the complete set of recognized interests as divided into two bundles. One bundle belongs to a private party or group, and the other bundle belongs to the government. The large number of sticks allocated to private individuals and groups suggests that private parties dominate determinations of land use.

## 3.1.2 COLONIAL NORTH AMERICA

The English attitudes and practices regarding land rights, restrictions, and responsibilities were brought to North America. They were adapted to the material circumstances of North America (Cronon 1983). An important physical condition was the vastness of the new continent—land abundance rather than scarcity.

An initial system of communal land interests in the Plymouth colony was quickly abandoned in 1623 in favor of private land interests in specific parcels (Bradford 1948). The community "had decided that each household should be assigned its own plot . . . the Pilgrims had stumbled on the power of capitalism. The Plymouth colonists never suffered a fear of starvation after this" (Philbrick 2006).

The colonists' isolation from close scrutiny by officials in England gave impetus to independent actions, a need for innovation, and a sense of empowerment among the colonists. In

Massachusetts Bay, all feudal restrictions on land use were ended in 1641 because "freehold tenure of lands . . . seemed in accord with the laws of God" (Morgan 2006).[1]

## 3.2   Postcolonial America

Early postcolonial America was largely a rural, agricultural society. Industrial activity was limited and in its earliest stages. The first US Census in 1790 indicated that 98 percent of the American population lived in rural areas.

"In Western societies the first step toward control of an environment usually is the assigning of tracts of land as grants of property—done by drawing lines on paper, although little may be known about the tract that is to be colonized" (Johnson 1976, 21). "The desire for assured ownership of separate property parcels required simple and accurate descriptions of landholdings, and the Ordinance of 1785 was a plan in support of such possessive individualism" (36).

The Ordinance of 1785 was the first legislative act of many acts designed to assure—in the words of Thomas Jefferson—that "as few as possible shall be without a little portion of land" (Johnson 1976, 39). The American legal historian, J. Willard Hurst, described the American leaders of that generation as seeking the goal of "diffusing political power by diffusing economic power" (Hurst 1975).[2] Quick survey of the public lands and their sale at low prices or by free grant were widely supported policies.

Large landowners understood the economy, social status, and wealth in terms of land ownership. The poor and immigrants saw their economic and social status as workers and craftsmen in their relationship with land and property. Land ownership and status were intimately related. "Real property meant more than land: the term applied to that cluster of privileges and rights which centered on land, or on the exercise of power which had a focus in some point in space" (Friedman 1985, 230).

"After 1787, the vast stock of public land was at once a problem and a great opportunity. . . . American society faced a central issue: in what way to map, settle, and distribute this almost limitless treasure of land." (Friedman 1985, 230–31)

The answer to this issue was the Public Land Survey System (PLSS).

"Once land was surveyed, it was disposed of. The government did not choose to manage its land as a capital asset, but to get rid of it in an orderly way. . . . On the surface, one

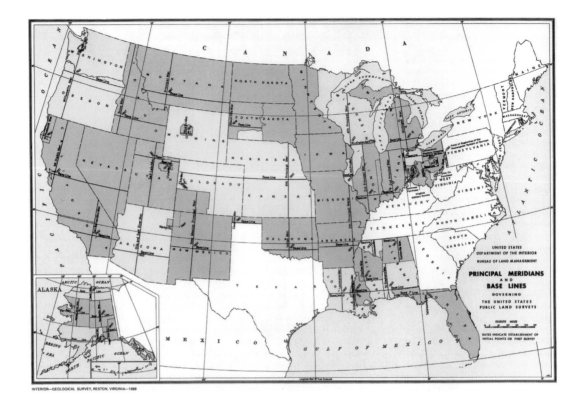

**Figure 3.2** The spatial extent of the PLSS.

Courtesy of the US Geological Survey and Bureau of Land Management.

sees in this policy the powerful influence of free enterprise and laissez-faire. . . . [D]ivestment was the goal, yet never the whole goal, or the policy. The reason for getting rid of the land was as a means of strengthening the dominant form of land tenure, and affirming the basic postulate of American social structure. The ideal was a country of free citizens, small holders living on their own bits of land." (Friedman 1985, 231–32)

The spatial extent of the PLSS is represented in figure 3.2.

The bundle of rights that a landholder received from the government under the PLSS process is that represented on the left of figure 3.1. The rights reserved to the government, depicted on the right, were recognized but rarely exercised under the frontier conditions of distant, poorly organized governments. It is not surprising that many landholders came to believe, "It's my land and I can do what I want with it."

## 3.3   The impact of nineteenth-century industrial development

The Industrial Revolution added new attitudes and practices to the American land concept without eliminating significant, traditional practices.

The retained, traditional land concept is represented by the actions of settlers in Pike Creek on the southeastern Wisconsin shore of Lake Michigan in 1836. The settlers established themselves on the land before the government had taken the surveying and other actions prior to land grants. They were aware that "from the survey ordinance of 1785 on, squatters settled large areas of the public lands in defiance of law ahead of official survey, without color of title other than that created by the impact of a popular feeling that would not be denied." These settlers wrote a document that stated their belief that the "Government has heretofore encouraged that our settling and cultivating the public lands is in accordance with the best wishes of the Government" (Hurst 1956).

As the Industrial Revolution progressed, land also became a platform for factories, canals, railroads, urban activities, and other new commercial and industrial activities. Transfers of land occurred more frequently than in an agrarian society where land often remained in a family for generations.

The early period of industrialization enhanced the American interest in turning land into a more readily transferable good. The land ordinance was actively used to reduce restrictions that inhibited transfers of real estate. Property was seen to be improved by its passing from hand to hand (Hurst 1956, 13). This represented a "preference for dynamic rather than static property, or for property put to creative new use rather than property content with what it is" (Hurst 1956, 28). "Land could not only be traded on the market: it was traded, openly and often" (Friedman 1985, 235).

Rapid land transfers made it easier to estimate the value of one parcel by comparison with another similar, recently sold parcel. The ability to estimate the value of land as a commodity made it not only possible but practical to operate a property tax system at the local level of government. Local assessors could make better, more easily defended estimates of land values because of the more frequent land sales. Local assessments sustained local property taxes and local government itself. The assessment office became an important local government institution.

The dominant midcentury preference was for property as an institution of growth rather than of security (Hurst 1956, 28). The disposal of public lands was a demonstration of the priority assigned at the time to immediate growth over the wealth of the future (67).

Land continued to be sought for traditional farms with traditional land and water rights, interests, and responsibilities until settlers encountered the dry conditions of the arid plains. Then changes in attitudes and practices led to new institutions and new land laws.[3]

# 3.4 After the American Civil War

The American legal historian Lawrence Friedman noted that "public lands continued to be a major topic of controversy in the second half of the century. The idea that the great federal treasury of land should be used to raise money was all but dead. The clamor for free land reached its climax in the famous Homestead Act of 1862; and government continued to use land heavily as a kind of subsidy. The Morrill Act gave away a vast tract of land to the states. . . . The land was to be used to establish 'Colleges for the Benefit of Agriculture and Mechanic Arts.'" (Friedman 1985, 414–15)

After the American Civil War, the federal government used its land resources to encourage development in the West. Public support for transportation shifted from turnpikes to canals and rivers and then to railroads. People wanted government to help mobilize land and resources to build transportation facilities (415).

"Toward the end of the century, the change in national attitude associated with the death of the frontier began to make its mark on public-land law. The underlying trait of policy, in the beginning, had been a kind of roaring optimism. . . . In the last years of the 19th century, the psychological horizon darkened. Out of a new sense of scarcity, and muted pessimism, the seeds of the conservation movement grew" (419).

After about 1870, a new order emerged, with the following four aspects:

1. Big industry

2. Big finance

3. Continued large population growth with a relative growth of cities

4. Increasing interdependence of activities (Hurst 1956, 71)

However, popular expectations of a rising material standard of living based on industry brought a new sense of helplessness in the promising society (73).

## 3.5   Science and changing land concepts

Faith in the ability to master the environment was a powerful force at the time. Science and technology offered the prospect of more and greater comfort. Throughout most of the nineteenth century, governments sold or granted their land without adequate regard for special resources in soils, water, or minerals that would distinguish one parcel from another.

The story of John Wesley Powell illustrates a relationship between developing and accelerating knowledge about the natural world and the slowly changing concept of land in America. Powell became a national hero after his voyages down the Colorado River

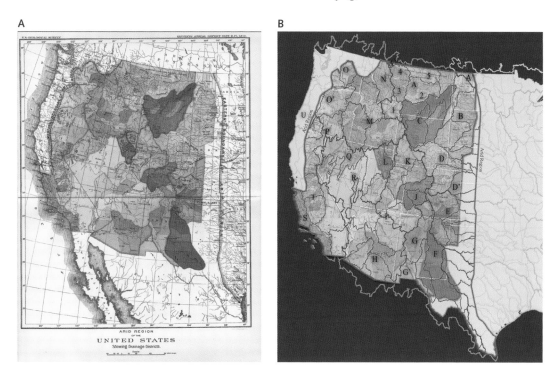

**Figure 3.3** In 1890, John Wesley Powell conducted a survey by horseback of the arid lands of the western United States (A). When the survey is compared with current high-resolution watershed boundary datasets (B) by the US Geological Survey (USGS), Powell's map is proven to be remarkably accurate. As director of the USGS in 1891, Powell called for a land management arrangement that would better utilize water resources in the arid west.

*Esri Map Book, Volume 27* (Redlands, CA: Esri Press, 2012), 62; courtesy of the US Geological Survey.

in 1869 and 1871. His scientific training as a professor of geology at Illinois Wesleyan University directed his observations of the region. He concluded from these data that western water supplies could at best provide irrigation to only 3 to 5 percent of the lands west of the one-hundredth meridian, that the transport of water from one watershed to another should not be allowed because that would leave the origin watershed with severe water shortages, and that state boundaries in the region should follow watershed boundaries in order to avoid interstate water disagreements.

Powell's ideas were seen as contrary to the prevailing land concepts and to the institutions that had been a part of American society since its origins, and his ideas were eventually rejected. Powell's scientific observations and conclusions were overcome by the prevailing land concept.

Natural sciences progressed to the point where they generated challenges to traditional understandings of the natural world. Modern ecology and the science of ecosystems began in this period. The mysteries of nature came to be understood, or better understood, as natural phenomena. However, the pace of advances in scientific knowledge was not accompanied by the same degree of change in the land concept and land institutions. For example, hydrogeology began to unravel some of the mysteries of natural water systems. Although it was possible to connect the hydrogeology of ground and surface water systems, water rights law had long separated surface and ground water legal regimes. The ability to connect the legal regimes for surface and groundwater rights remains a work in progress (Getches 2009).

## 3.6    Twentieth-century land concepts

By the beginning of the twentieth century, the new science and technology made it possible to measure and understand land in ways that were previously unknown. Attitudes and practices, including laws, began to change, however slowly.

Consider changes in what are now called wetlands. These were swamps in the nineteenth century. Swamps were wet areas to be drained so that the land could become useful farms, residences, and factories. Water was legally a "common enemy" in some states (Getches 2009). Twentieth-century science revealed that these areas are useful for wildlife habitat, waterway purification, flood mitigation, stormwater barriers, and aesthetic values. Citizen interest groups emerged as early as 1937 seeking to retain or restore these areas. Ducks Unlimited was formed and began to purchase and protect lowland areas that sustain waterfowl habitat. At the same time, drained wetlands remained vital to food production. They have been described reverentially in Jane Smiley's (1991) Pulitzer Prize–winning novel, *A Thousand Acres,* as "a farmer's patrimony." The traditional land and resource attitudes and practices do not always yield to the new. The old and the new attitudes and practices exist in a complex relationship.

Recent events reveal the folly in a continued, significant separation between what is known scientifically about land and its resources and their uses. The human and physical devastation from Hurricane Katrina in New Orleans was due in part to loss of protective coastal wetlands. Appropriate information about the location and impact of wetlands was not adequately integrated into planning and management of their uses.

In the twentieth century, population and urban growth, the growth of big industry and finance, and the development of the science of natural areas altered the American land concept, increasing conflict over land use. Previously, the dominant practice of privately determined land use was expressed and documented in private land-use transfers, covenants, and contracts. Copies of these documents were deposited in register of deeds offices that were designated to give notice of the transaction to others in the local community. Private organizations and individuals searched these registries for these documents. These parties provided private analysis and assessment of the documents' meaning regarding the nature and extent of privately arranged distribution of land interests.

The increased nature and level of land-related activity in the twentieth century generated increased conflict over land and resource use. A partial solution was found in the increased use of the government powers to take and control land use, which are illustrated on the right side of figure 3.1.

"In land law, it was a century of land-use controls. . . . [Z]oning became an almost universal feature of the land-use laws of cities. Planning and controls were used to monitor the growth of the city, to preserve the character of neighborhoods, to stop any downward slide of land values, to counterbalance the iron laws of the market. These were popular goals, and zoning served the interests of the middle-class homeowner, and (to some extent) the businessman too" (Friedman 1985, 678). These goals were controversial too, allegedly a violation of the Fifth Amendment to the US Constitution's prohibition of taking land except for a public use. However, land-use control ordinances received constitutional approval by the US Supreme Court by the middle 1920s (Pennsylvania Coal Company v. McMahon 1922; Euclid v. Ambler 1926).

Public land-use controls to protect health and safety and property values proliferated. "Air and water pollution, urban squeeze, and other symptoms of distress were old enough; there had always been voices crying in and for the wilderness. But the voices became more strident from the 1960s on. A real sense of doom began to hang over this small and limited world. Many sources fed the ecological movement. First, economic growth . . . no longer satisfied everyone. . . . Second, the crisis was real. Resources were not infinite. Big business was poisoning the rivers and darkening the air. . . . Third, this was a society with many rich and leisured people; with enormous government and governments, peopled by professionals and bureaucrats; and with a growing number of academics and intellectuals looking for a place in the sun" (Friedman 1985, 680).

Advances in natural resource and environmental science led to a greater understanding of the functions of natural systems. The transformation of a nineteenth-century swamp into a twentieth-century wetland is an example. Although the functions of natural resources are better understood, the understandings and attitudes are not as fully incorporated into the land laws and processes as are the private contracts and some public land-use controls.

The ability to incorporate new understandings and attitudes of the natural world into land law and process is limited because some pollution problems can be seen clearly, while others cannot. For example, the Cuyahoga River in Cleveland caught fire in 1969, a river turned orange during the night when a large truck carrying waste passed through town at midnight, and the air turned brown over Los Angeles. These observations contributed to support for passage of air, water, and other pollution-control laws in the 1970s. At present, the cause and impact of global warming and similar issues cannot be seen in the same way.

In the 1970s, Congress passed and the president signed a series of environmental laws. These included the Clean Air Act (CAA); the Clean Water Act (CWA); the Federal Water Pollution Control Act (FWPCA); the National Environmental Policy Act (NEPA); the Toxic Substances Control Act (TOSCA); the Resource Conservation and Recovery Act (RCRA); the Comprehensive Environmental Response, Compensation, and Liability Act (CERCLA, also called "Superfund"); the Coastal Zone Management Act (CZMA); and many others. Although these environmental laws have been amended, the laws passed in that period established a framework of rules, procedures, and practices that characterize the current operation of the laws.[4]

One aspect of these federal laws is that they regulate the operation of facilities but generally leave determination of the location of facilities to local and state governments. For example, the CAA gives the federal government the power to determine the allowed concentrations of designated air pollutants (the criteria pollutants) for all regions of the nation. The CAA also gives the federal government the power to determine the upper limit of the concentration of pollutants from facilities. Although the federal government determines the overall quality of the air and what is emitted from facilities, the CAA leaves to the states the task of determining how the standards are met. This task allows state and local governments to retain the power to determine the location of each facility as a part of the state implementation planning and management.

A related aspect of these environmental laws of the 1970s is a law that was talked about but not passed. This law would have authorized the federal government to control use of land in some circumstances and led to federal land-use control legislation like that commonly used by local governments.

The relevance of these actions and inactions regarding environmental legislation indicated the ongoing strength and importance of traditional land-use attitudes and actions. The traditional emphasis on private land-use determinations was retained. A recognized need for collective action, in the form of environmental laws, left most land-use-control legislation, with its attention to facility siting, as close as possible to the individual, mostly at the local government level. These preferences, whatever the balance of their advantages or disadvantages, are strongly held by Americans. This is a powerful message to those who would change land-use law and land records.

The use of government powers to affect land and resource use has significantly increased in the last century. This dynamic process continues, resulting in a major change in the understood relationship between private and public land interests. As depicted in figure 3.1, the two sets of interests were viewed separately. In the nineteenth century and even later, an American might have said, "It's my land, and I can do what I want with it." That statement was never correct legally because of the government powers at that time. It would be appropriate to say that the nature and scope of exercised government powers were limited or not implemented in all areas.

At present, the considerable increase and use of government powers to influence land and resource use results in a more complex relationship between public and private land interests.

This relationship is depicted in figure 3.4.

**Figure 3.4** Overriding public land interests affecting the exercise of traditional private land interests represented in figure 3.1.

Figure 3.4 illustrates the large number of public land interests exercised in twentieth- and twenty-first-century America. The result is a complex, overlapping or simultaneous set of private and public land interests. Land can now be described as a "commodity affected with a public interest" (Babcock and Feurer 1979).

The nature and extent of exercised government powers over use of land and its resources constitute a set of overriding public interests. The traditional public actions in taxing, taking, and controlling land interests, simply represented by single sticks in figure 3.1, have become a large array of exercised public interests in twenty-first-century America, as represented in figure 3.4. This development has, not surprisingly, generated considerable dispute about the meaning and limits of exercised public interests. The resulting court decisions and interpretations must be consulted in order to fully understand the nature and scope of many of the exercised public interests.

Private rights or interests are retained subject to any overriding public interests. For example, although a landowner retains the private right to subdivide a parcel, many communities have overriding subdivision ordinances that determine the conditions for that subdivision (e.g., resulting parcel size). Similarly, although a landowner retains the right to drain a wetland, overriding legislation and administration often exists that determines the conditions under which the action is allowed.

The nature and extent of exercised government powers results in many disputes that test the nature and extent of government authority. These disputes often reach the courts where judicial opinions establish overriding government interests. For example, a landowner retains the right to establish covenants in a deed that restrict how a buyer uses the land. This can take the form of a covenant restricting subsequent conveyance of the land to a non-Caucasian. The overriding public interests expressed in the form of a US Supreme Court decision that prohibits governments at all level from any enforcement actions that sustain the covenant (Shelley v. Kraemer 1948). Another example of the use of judicial opinions in court cases to create overriding public interests in land occurs when courts limit agency discretion in the implementation of land-use-control legislation. For example, this occurs when a court finds that a required ten-to-one replacement of drained wetlands, a form of impact fee for the permit to drain, is excessive (Koontz v. St. Johns River Water Management District 2013).

The existence, location, and substance of these overriding public interests in land and resource use and their connection to parcels and areas are in a state of arrested development. Records and information are scattered among many agencies and courts. This information is not regularly gathered and indexed in a central registry in the way that records of private interests are gathered and indexed in the register of deeds. It is difficult for a landowner or the community to know what can be done with a parcel or area when all the public and private interests are considered. Uncertainty affects action and inaction with economic and social impacts. Those with the resources to locate, gather, and use the scattered set of records are given an advantage in planning and management. Modern technology makes it efficient, effective, and equitable to establish a central public registry for many if not all of these overriding public interests in land and resources.

The role of well-defined property rights in the determination of land and resource use was considered by the economist Ronald Coase (1960) in one of the articles cited as the basis for his Nobel Prize in 1991.[5] He challenged the prevailing concept that the only way to restrain people and organizations from harming others, as occurs in environmental injuries, was by means of government regulation. He noted that, where appropriate and practical, parties often make privately negotiated arrangements to resolve conflicts, taking into account the existing law but working around its provisions to achieve mutually satisfactory outcomes. This result is encouraged by a precise definition of rights, reduction of transaction costs attendant to negotiations, and access to information by all parties. Clearly, a modern land records system contributes to this goal of negotiated resolution of conflict.

It should be noted at this point that the discussion in this chapter has not given the history of environmentalism over the last 150 years as much attention as that subject often receives. This exclusion has been done deliberately for two reasons: First, the history of environmentalism is most often taught and discussed. Second, and most important, examination of the longer land history reveals why traditional thoughts continue to be so strongly held and acted on. For a summary of the history of environmentalism over the last 150 years, see *The Environmental Policy Paradox* (Smith 2009).

## 3.7    Conclusion

Americans' attitudes and practices regarding land status and interests are the complex result of historical development. On the one hand, there is the preference by the private landowner for the right to use land and its resources as the owner desires. On the other, there is the collective desire by citizens to have some power through government to control land use for a public purpose based on common interests. "One theme stands out, then, in the tangled history of American land law: private ownership, and not by a small elite, but by millions of people" (Friedman 1985, 49).

The following land designations highlight the complexity of human understanding of land status and thereby provide a context for this land-use drama:

- Land as private property. Private property refers to a set of recognized rights held by someone who has the power to dispose of a thing in every legal way, to possess it, to use it, and to exclude everyone else from interfering with it. Land as private property conjures the image of an entity divided into parcels, lots, tracts, and so on, that can be bought, transferred, leased, and sold.

- Land as public property. Land can be owned and managed by a government. Everyone in the community may use this land, which includes parks, schools, roads, lakes and rivers, and so on.

- Land as a common pool resource. This is a resource that can be used by many without subtracting from the ability of others to use it in the same way (e.g., a weather report can be used by many without interfering with others' ability to use it similarly), and it is hard to exclude many from its use. A body of water is a common pool resource when pollutants are placed there. This is a community resource subject to various forms of community management (Ostrom 1990, 2005).

- Land as a natural resource. A natural resource is a material removed from the earth for human use. It also is the natural system that supports life on a continuing, sustaining basis—one that needs protection.

- Land as a cultural resource. People apply important cultural associations to land. Concepts of home; sense of place; and religious, historical, and aesthetic qualities are attached to specific locations. Many of these elements are unseen. The importance of unseen, not easily measured cultural aspects has long been recognized. "Things which are seen are temporary, but the things which are not seen are eternal" (2 Cor. 4:18, New King James Version).

- Land as a platform for activity. Modern urban and rural American cultures lead people to think of land as the platform for life's activities. Urban property is a city lot or suburban tract with streets, houses, driveways, stores, and sewers. Rural property is a tract with forests, creeks, interstate highways, and pipelines. These land concepts can separate people from awareness of, and differences between, the urban needs and land's natural elements.

A particular American historical development has led to these attitudes and practices. They form a complex, inconsistent set. Nevertheless, the need to make wise land- and resource-use determinations in the twenty-first century makes it imperative that an effort be made to integrate these perceptions. Land records and information institutions must reflect this need. Design and implementation of these institutions depends on identifying the appropriate land records and information that represent these needs, connecting these materials, and distributing that material to all the parties interested in land planning and management.

Perhaps, the motto in this modern technological age is this: "Mold the technology to fit that which we hold dear."

These are the perceptions that the earliest English settlers in America brought with them, and how they developed over the centuries. There is still work to be done to make modern technology fit that which we hold dear regarding land and its resources.

# Notes

1. We have relied on the words of the eminent colonial historian Edmund Morgan, who has been a professor of history at Yale for thirty years and is the recent author of the *New York Times* best-selling book, *Benjamin Franklin*. He considered the tension between individual freedom and the demands of authority in the story of the Puritans who confronted this dilemma early in American national development (Morgan 2006).

2. The concept of the yeoman farmer as the basis for an egalitarian, American democracy is sometimes viewed as a narrow, antiquated, long-lost Jeffersonian ideal. However, the concept can be seen metaphorically (Hurst 1956, 1975).

3. The settlement of the arid western plains of North America has been described as an institutional fault line in American land planning and management. Faced with water conditions unlike any that were a part of English, English colonial, and American experience, those Americans who experienced the conditions initiated and guided the development of new water rights laws that reflected what they felt they needed. The result is a pattern of water laws in the United States that still largely divides the states into two water law institutions (Webb 1959). For a description of modern water law, see Tarlock et al., 2002.

4. Congress passed many important environmental laws in the 1960s and 1970s. This great period of environmental legislation has not been repeated since that time. The structure and form of these laws still dominate modern environmental law after decades of revision and amendment.

5. The Nobel Committee cited an article written by Coase (1960). Coase's ideas have assumed a significant role in environmental law and economics. In his textbook on environmental law, Professor J-M Stensvaag begins his discussion of Coase's ideas with the statement, "The Coase analysis is fundamental to an understanding of modern environmental law" (Stensvaag 1999, 60).

# Chapter 4
## Land records in America

The history of land records systems in America is related to the history of land. The history does have aspects of its own that bear close examination for information relevant to change in the system. Of special interest are those aspects that indicate how the developed system serves the demand for data and information in land planning and management.

The conditions of early English settlements isolated from England and each other favored the development of a land transfer or conveyancing process dominated by private-sector actors. Private professionals helped with negotiations, prepared documents, located documentation of previous transfers, interpreted and summarized the meaning of documents, examined descriptions of the location and extent of land interests, identified and marked the location and extent of interests on the ground, and provided assurances to the purchaser.

The tasks were divided among several private professions. Abstractors located documents and summarized their contents. Surveyors located, marked, and described boundaries. Lawyers interpreted the meaning of existing documents, wrote transfer documents, and gave assurances to buyers. The private professionals established practices among their members that became the legal and professional standards for their work.

## 4.1   Land records in colonial America

"The early North American [records] arrangements were designed to promote quick, efficient, and secure land settlement" (National Research Council [NRC] 1980). The conveyance of ownership of public land to private individuals and groups was a means to induce European immigration. Some landowners and groups (Lord Ashley et al. in

South Carolina, Lord Baltimore in Maryland, William Penn in Pennsylvania, James Oglethorpe in Georgia, etc.) received large land grants from the Crown in a manner reflecting residual feudal practices. These landowners and groups used conveyance of some of their land to induce immigration.

The granting of land, including the description and marking of boundaries, was initially the responsibility of the colonial surveyor general. Some grants of land were made without boundary marking. A smaller parcel created from the larger ones was sometimes surveyed and boundaries were maintained by a private surveyor, the purchaser, or not at all (NRC 1980).

Parties were free to describe the land interests transferred, to create new types of interests (limited only by land interests expressly forbidden by law), and to describe the location and extent of interest boundaries, all without involvement of a government official. The metes and bounds form of boundary description was widely used in colonial America. Sometimes, this informal system worked well; other times, chaos was the result. There was a need to provide some order to the chaotic process of land conveyance, especially in the frontier setting where security of ownership was always an issue.

As indicated in chapter 3, land could not only be traded, it was traded openly and often. One aim of land law in the new context was to keep the land market open and mobile in a society where there were few trained lawyers, surveyors, and other professionals. Government offices and officials were often few and far from citizens, especially in the rural areas. Changes in land law were, initially, empirical (Cronon 1983). Conveyancing law governing the transfers of land interests and their documentation had to be changed. Sustenance of the land transfer market required that land documents become simple and standard. In early America, every man (and later, every woman) was or might be a conveyancer (Friedman 1985, 236).

The result was a colonial land record system that combined old English private conveyancing practices with a government recording system that was minimally involved in substantive details of the land interests transferred. Government offices (e.g., registers of deeds) were established, often at the county or town level, where copies of transfer documents could be deposited and stored. The register was authorized to receive the document and to identify but not validate the stated names of the land interest grantor and grantee and the stated description. No government authority was given to claims in the document. Assessment and assurance of the nature or validity of interests allegedly conveyed in a document was a task for private professionals, if done at all.

Recording of land interest transfer documents in a public office without government assessment of document claims is an American innovation, reportedly established in the

Plymouth colony around 1624 (Shick and Plotkin 1978). This innovation closely followed the Plymouth colony's introduction of privately held land parcels in 1623 (Philbrick 2006; Deetz and Deetz 2000).

Four characteristics were found in colonial statutes establishing the government land-interest-transfer recording offices:

1. The instrument of transfer—such as a deed, mortgage, easement, or other contractual arrangement between a buyer and seller—must be acknowledged by a public official before the document is received and recorded.

2. The entire instrument must be recorded.

3. Legal priority generally is assured to the grantee by the act of recording.

4. The instrument of transfer is operative without recording, with the interest passing at the time of execution and before it is recorded (NRC 1980).

The government provided a known, secure repository for copies of the transfer documents. Receipt of documents by the register of deeds was authoritative, even though there was no government analysis of the authenticity of claims in the documents. Recordation was a means to give legally recognized, authoritative notice to the community of the existence of the transfer document. All parties to subsequent transactions were legally responsible for knowledge of the existence of the recorded documentation of earlier transactions. The government did not examine or give consideration to the validity of declared or implied claims in the documents. Determination and assurance of the validity of the documents and the ownership of land interests generally was left to private individuals and professionals. These principles and practices dominate the existing American land titles system.

## 4.2  Postcolonial America

Several forces influenced the design and practice of land records institutions in early, postcolonial America. One was the desire to avoid the unsystematic, chaotic surveys and the unregulated private description of rights and boundaries in colonial America. Another was the development of the science and practice of land measurement—surveying. A significant number of private surveyors could now execute tasks previously limited to government surveyors. Finally, there was a desire to transfer the public lands west of the

Allegheny Mountains into private ownership in a manner that promoted security of title and boundary.

Systematic observations of land conditions, measurement and marking of parcel boundaries, and careful documentation of these efforts were to be made by federal surveyors before conveyance of public land. The result was the Public Land Survey System (PLSS), which was first elaborated in the Ordinance of 1785.[1]

The PLSS can be viewed in several ways. First, it had political and economic objectives and outcomes. These objectives and outcomes led to the distribution of land to many landowners rather than a few large landholders as part of an effort to distribute political power by distributing economic power to many. Second, the PLSS sought to overcome some of the difficulties attributable to the earlier, unsystematic survey practices. Previously, government surveyors measured and marked the boundaries of large parcels granted by the Crown, while private parties surveyed subsequent divisions of these large parcels. Third, the PLSS created a public record of measurements and markings of the parcel boundaries created by the government prior to the government's conveyance of land. Fourth, the federal surveyors gathered and reported on information about the condition of land and water near the boundary monuments they placed in the ground.

Collection and documentation of information about the condition of the land and its resources were inseparable from the land measurements and boundary markings, gaining the PLSS the attention of those who sought to take and develop the best lands. These records and information about the conditions at the time of settlement often remain important today in land-use determinations, representing valuable and often the only descriptions of land conditions at the time before intensive and extensive changes to the land. The PLSS provided a basis for development of land, government, and education in the western lands. The PLSS was a major implementing mechanism for a grand land planning and management process for the extensive federal lands in the West.

The PLSS provided security of ownership and boundary in frontier circumstances. The PLSS provided a basis for achieving personal and community objectives regarding land and its resources. The system relied on actions by the federal, state, and local governments with which citizens and officials felt comfortable and which they supported. The system gave individuals and groups a means to achieve what they wanted: it left them in charge of much of the land development. Local recordation offices, such as the register of deeds, dominated the land records and information institution after the federal officials moved on.

The PLSS was applied to most but not all land west of the Allegheny Mountains. This application left many areas in the eastern United States outside of the domain of the PLSS. However, the example and value of combining records of the nature and extent of land interests with observations of land conditions appropriate for land planning and management, which was essential to the original PLSS activity, is a lesson for development of future land records systems both within and outside of the PLSS areas.

## 4.3  Impact of the Industrial Revolution on land records

The new uses for land and its resources generated by the Industrial Revolution accelerated the American propensity to transfer land and move on. Tocqueville observed that "an American farmer . . . especially in the . . . West . . . brings land into tillage in order to sell again . . . on the speculation that, as the state of the country will soon be changed by the increase of population, a good price may be obtained for it" (Tocqueville 1945, 157).

The increased rate of land transfers made it easier to estimate the value of a parcel by reference to the price of a recently sold, comparable parcel. This ability meant that the value of land and its improvements could be estimated reasonably well by local citizens, local professionals, and local officials.

The local assessor's office was developed in this context. The office established and sustained a locally operated property tax system that provided the major source of revenues for local government. The office became a powerful, local government institution whose job was to assign a defendable value to land and its improvements so that a property tax bill could be sent to the owner with the reasonable expectation that the tax would soon be paid.

The statutory duty of these assessing offices is to value parcels and assign a property tax. The parcel map was developed as a means to make these tasks efficient. The local government assessor or auditor became the source of data and information about land parcels, especially those data needed for estimating land values, and became the local property parcel mapmaker. The assessor is now the de facto municipal spatial data manager in many jurisdictions. This role has continued to expand with the development of geospatial technology, and the assessor is one of the key geographic data managers in local government.

Assessors' enabling statutes rarely mentioned, much less defined, the quality of parcel maps. The parcel map acquired no legal authority, even for land values. The assessor's legal duty was to provide a set of defendable parcel valuations. The parcel map's attributes could be argued on a case-by-case basis. These characteristics, developed in the nineteenth century, apply to local parcel maps in most existing assessing offices in the twenty-first century.

# 4.4    After the Civil War

The accelerating industrial and commercial activity after the American Civil War revealed difficulties with the existing deeds recordation system. Documents relating to a particular parcel had to be located from among the large and growing set of documents in the register of deeds. Recordation was not required by law in order to transfer rights, despite the offer of public notice. Transfers occurred when the parties exchanged transfer documents and compensation, not when a document was presented to the register of deeds. Intergenerational transfers sometimes were not documented and recorded until the time of transfer out of the family. Parties to a transfer retained the freedom to define and describe the nature and extent of transferred rights. A new type of land interest (e.g., time sharing) could be created as long as it was not a property right expressly contrary to law (e.g., a discriminatory clause). Parties retained the freedom to describe the parcel's location in a manner of their choice as long as the parcel could be unambiguously identified. Completeness, timeliness, uniqueness, and other factors became land records issues.

These problems were manageable as long as land planning and management remained largely local, vernacular matters. After the Civil War, big industry and big finance became increasingly national. Demands grew for transfer and assurance practices that satisfied people far removed from the parcels. Indexes to documents (i.e., grantor/grantee and tract indexes) were introduced in the register of deeds offices. At the same time, private-sector activity developed to sustain and promote land transfer and property tax actions, to satisfy new national demands, and to provide for entrepreneurial businesses that satisfied public and private demands for land data and information. Appropriate documents had to be located, and their meaning had to be determined quickly and given sufficient credibility such that interested parties could take desired action in reasonable time. Those who searched the records, provided opinions about the status of title and boundary, and insured those opinions—professionals such as abstractors, title attorneys, title insurance representatives, and those who prepared and sold parcel maps—developed and expanded their services to a private market. Symbiotic public and private partnerships developed between the public offices and the private conveyancing professionals and organizations. Private companies generated parcel maps in response to specific private and public demands.

Title insurance by private companies was developed in the nineteenth century. These companies went beyond the provision of insurance for title attorneys. They provided opinions and guarantees to purchasers of land and those interested in the status of ownership. By the end of the century, practices, and laws reflecting these practices, were standardized operations and represented an institution in the community.

In the latter half of the nineteenth century, the Australian civil servant Robert Torrens devised a system for government, rather than private, determination and assurance of the status of parcel land interests. Canada successfully adopted this type of system in some of its western provinces.[2] However, a Torrens-type title records system was not to be the American way. Efforts to introduce the system were made after the economic depression that began in 1893. About a dozen states adopted enabling legislation. However, these efforts were not successful because they failed to replace the established, private conveyancing practices. The American legal historian Lawrence Friedman noted that "as of 1900, then, title registration was at best a hope, at worst a missed opportunity" (Friedman 1985, 435).

An effort to promote a Torrens-like system in the United States occurred during the Great Depression of the 1930s. This effort was associated with problems resulting from the title records chaos following large numbers of mortgage foreclosures and land abandonment. This effort to expand Torrens activity was largely unsuccessful (Shick and Plotkin 1978).

A title registration system was considered in the late 1970s when mortgage interest and inflation rates were extraordinarily high. The Real Estate Settlements and Procedures Act of 1978 (RESPA) called for a study of real estate settlement costs by the US Department of Housing and Urban Development (HUD). This review included a specific study of a title registration system in the United States.[3]

The system of private-sector abstractors, title attorneys, title insurers, surveyors, private real estate mapmakers, and other private land information entrepreneurs in a symbiotic public-private partnership with public offices, such as the registers of deeds, was established by the end of the nineteenth century. Title companies often copied every document deposited in the register of deeds, indexed these documents, and provided access to them for a fee. Some collected and indexed abstracts, which represented a form of outsourcing public records, an activity further developed in recent times by those who handle mortgages.

The fundamental roles of the actors and organizations were sustained throughout the twentieth century and continued into the twenty-first century. The system of title records in America reflects a set of powerful, established institutions that sustain the conveyancing process and form a barrier to changes that serve land planning and management.

## 4.5   The influence of twentieth-century science

Changes in the understanding and knowledge of the natural world and of natural areas altered the types of land data and information collected and used for land planning and management. When wet areas are seen as an enemy and called swamps, one type of land data and information is needed. When they are seen as a wildlife habitat to be sustained as wetlands, then other data are needed.

Land measurement sciences and engineering—including surveying, geodesy, photogrammetry, cartography, and remote sensing of the environment—significantly changed the ability (accuracy, speed, cost, etc.) to observe, measure, and represent the land features and their locations.

Technologies other than those associated with land measurement affected the traditional activities by the register of deeds, assessor, and the related private professionals. New technologies, such as photocopiers and printers, made it possible and practical for private title records companies to collect copies of land records in the register of deeds and create their land title records systems. These technologies improved the efficiency of the traditional title examination and assurance process without disturbing or expanding the nature and scope of the institution. Little change was made in the process of connecting records of privately and publicly established land interests.

Land-use control statutes were often passed on the basis of the results of new science and technology. For example, ecological science generated a change in land concept that converted swamps into wetlands with attendant changes in land use and land-use laws. Rules were promulgated, and permits were granted. These actions changed the nature of land interests not only at the parcel level but also at the level of land interests across the community. No organized system developed analogous to that for privately arranged interests in the land titles operations for locating, interpreting, and summarizing the nature and extent of land interests affected by the legislation, rules, and permits. Often, a landowner could get an answer to an inquiry about what the records in the register of deeds revealed and what data and information were used for the assessor's valuation. However, if the landowner inquired about what could be done with the parcel when all the publicly established land interests were considered, then a much more difficult undertaking was required.

Land-use development and control statutes in the twentieth century often came with demands for land data and information when a plan or permit was required before land uses were changed. Each had specific data and information requirements. A complex land-use change, such as a power plant development, often required several permits from different agencies, and each had its data and information requirements. The problems of gathering and arranging land records and information in order to obtain land-use permits have been described by a practicing information professional acting as an agent for the applicant (Daylor 1982).

Land records systems in the twentieth century operated with a distinction between types of records and information. Records that acquired a legally recognized status in the process of public land management were separated from those in the title examination process. Maps of land features generally did not acquire an authority in actions where they were considered for use in planning and management. These circumstances added to the general disorder and chaos associated with use of land data and records.

## 4.6   Land records in the late twentieth century

During the latter part of the twentieth century, there was renewed interest in improving the conditions and problems associated with land records and information. This renewed interest, called *modernization of land records*, accompanied the early development of modern, computer-based geospatial technology. Title and real property lawyers, land economists, land tenure academics, surveyors, engineers, landscape architects, and land and resources planners were among those who led this interest in reform of the institutional aspects of land records systems.[4]

In 1975, John D. McLaughlin became the first PhD in North America to address the subject of a modern land records system. McLaughlin's thesis was remarkable in several ways. First, it addressed many of the land records institutional problems described previously. Second, it came from a civil and surveying engineer well informed of the status of the emerging geospatial technology. Third, it introduced, or reintroduced, the term *cadastre* at a time when the term was unfamiliar to most Americans and largely confined to those who managed public lands in the American West, especially officials in the Bureau of Land Management (BLM).

A cadastre was defined as "a record of interests in land, encompassing both the nature and extent of these interests" (McLaughlin 1975). This definition focuses attention on both the location of land features and the nature of rights, restrictions, and responsibilities associated with these features. The breadth and scope of this attention derived, in part, from concerns with land records systems that served the demand for records and information needed for land planning and management. When the system does this, it becomes a multipurpose cadastre or multipurpose land records system.

The multipurpose cadastre or multipurpose land records system concept gives great emphasis to the role of parcel records and information as a window into a broad understanding of land and its resources and into the problems and issues associated with land planning and management in twenty-first-century America. The parcel is what people care about. It is the focus of individual and collective land planning and management activities that determine how land and its resources are used. The concept of a modern multipurpose cadastre, what we call a modern multipurpose land records system, retains the parcel as the basic building block for a modern system.

The McLaughlin thesis and its author greatly influenced the seminal 1980 report, *Need for a Multipurpose Cadastre,* generated by the NRC (1980) and published by the National Academy of Sciences. This report was the first of a NRC-generated series on various aspects of land records problems.

At the time of the 1980 NRC report, proposals to modernize land records systems emphasized the use of parcel maps, mostly in paper form. They also relied on a unique parcel number for each parcel, generated by reference to the parcel maps. This unique parcel identification number (PIN) could be attached to many land records that related to a parcel. The PIN was the practical basis for connecting the various types of land records and information with the parcels.

In the report *Need for a Multipurpose Cadastre* (NRC 1980), the parcel map is called the cadastral layer. This layer was the primary spatial representation of land. The concept of a multipurpose cadastre depicted in *Need for a Multipurpose Cadastre* is shown in figure 4.1.

Figure 4.1 depicts connections, or links, between parcel boundary maps and the locations of other land features. These links are built on an underlying foundation of geodetic

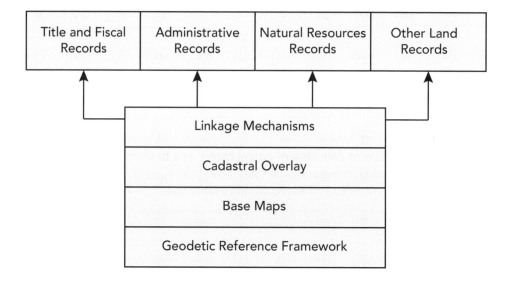

**Figure 4.1** Components of a multipurpose cadastre.

Figure 1.1 in *Need for a Multipurpose Cadastre,* (National Academies Press, 1980), 14; courtesy of John D. McLaughlin.

control information. This leads to the observation that geodetic control information means compatibility of information that is otherwise incompatible (Epstein and Duchesneau 1984). Maps of parcel boundaries and of the location of land interests in parcels, such as easements, rely on this geodetic control information. Connections or links to records that are not about the locations of parcels and other land features (i.e., textual records of the substance or nature of easements), use the parcel map to establish unique PINs. These PINs can be attached to all kinds of documents, records, and information that relate to a parcel and material about both the nature and extent of land features and interests. PINs allow for links between parcel maps in analog or digital form to the text of documents that describe the nature of property rights associated with parcels. These links make possible the combination of records and information about both the nature and extent of land features and interests for the purposes of addressing the full range of land-related problems.

Since 1980, geospatial technology development has emphasized the location of land features and interests. Connections to the nature of interests (rights, restrictions, and responsibilities) were not emphasized. PINs have not been fully deployed to make connections between the locations and nature of features and interests. This failure to fully use PINs has contributed to the slow establishment of a common parcel map across a state and across the nation.

It was reasonable in 1980 to use the paper parcel map as the basis for representing and connecting various documents and data about land and its features. Use of a paper parcel map as the medium for connecting many types of records meant that the typical scale of parcel maps became the de facto basis for the scale of the spatial unit used to represent many if not most other land features, not just the parcels. Maps of land feature locations (e.g., those for parcels, wetlands, building footprints, soils, and rivers) were made at or reduced to map scales common for parcel boundary maps. Data available at larger scales were often represented on maps at the smaller scales common for parcel maps, which often influenced the choice of pixel size in the development of computer-stored spatial databases.

The common scale of cadastral or parcel maps, as a basis for a multipurpose cadastre, was not a problem when making connections with title, assessment, or other textual records. However, representations of the location of land features—such as wetlands, soils, land cover, land use, roads, waterways, utility poles, and scenic views—could be made at larger scales, especially as geospatial technology improved. This meant that use of the cadastral parcel layer as a basis for a common representation of several features could result in generalization of larger-scale material to the smaller scale of the parcel map, with loss of important detail.

During the 1980s, the revolutionary changes in geospatial technologies altered the land feature location-mapping process. Global navigation satellite systems (GNSS), based on geodetic science and geodetic control information, combined with advances in electronic distance measurements, photogrammetry, and remote sensing of the environment to make low-cost, rapid, and accurate measurement of land-feature locations for almost all land features possible.

The number of those who could make these measurements and representations increased rapidly. The ability to do so came within the reach of many outside the agencies that traditionally had performed these activities. These activities were abetted by the rapid development of information technology and the ability to assemble and distribute the results to many. These activities also uncovered many errors and omissions in records and information that were not discoverable in the small-scale maps or without the map overlays made possible by the technology.

It was now possible to measure, represent on maps, and distribute data about the locations of land features without reference to a parcel map. Each feature location could be independently identified, measured, and described in a common geographic language, such as latitude/longitude and geographic coordinates without reference to a parcel map. Each land feature could be and was depicted independently. Each independent representation became a data layer that could be connected and combined with other layers using a common geographic language. This activity was enhanced and brought into wide use by the development of computer programs such as ArcGIS software, introduced in 1982 as ARC/INFO, and its successors. These programs made it possible to connect location-oriented data. Connections and combinations no longer depended on a common basemap, such as a cadastral or parcel map, but relied on spatial reference systems that were mathematically defined.

The ability to represent land features and make connections between independent representations of their locations is represented in figure 4.2.

The concept for a land information system represented in figure 4.2 goes beyond that of an abstract assembly of spatial data layers. This concept considers the complex set of public and private institutions in a local community among which the records and information are widely scattered. The records and information must be identified and located, and the institutions must be consulted and asked to contribute when a package of materials is assembled by an individual or organization for use in land planning and management.

For an extended discussion of the institutional aspects of figure 4.2, see chapter 4, "Data, Information Technology, and Institutions: Building Blocks for Land Planning and Management," in *Citizen Planners: Shaping Communities with Spatial Tools* (Niemann et al. 2010). A history of this diagram can be found in the article, "GIS Innovator: Innovation with Affect—Part Two" (Niemann and Niemann 1994).

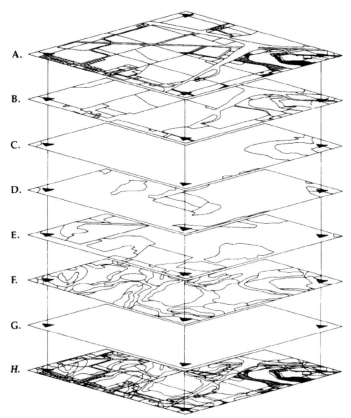

## Concept for a Multipurpose Land Information System

**Section 22, T8N, R9E, Town of Westport, Dane County, Wisconsin**

| Data Layers: | Responsible Agency: |
|---|---|
| A. Parcels | Surveyor, Dane County Land Regulation and Records Department |
| B. Zoning | Zoning Administrator, Dane County Land Regulation and Records Department |
| C. Floodplains | Zoning Administrator, Dane County Land Regulation and Records Department |
| D. Wetlands | Wisconsin Department of Natural Resources |
| E. Land Cover | Dane County Land Conservation Department |
| F. Soils | United States Department of Agriculture, Soil Conservation Service |
| G. Reference Framework | Public Land Survey System corners with geodetic coordinates |
| H. Composite Overlay | Layers Integrated as needed, example shows parcels, soils, and reference framework |

Land Information and Computer Graphics Facility
College of Agricultural and Life Sciences, School of Natural Resources

UNIVERSITY OF WISCONSIN-MADISON

**Figure 4.2** Concept for a multipurpose land information system.

Image generated by Sullivan, Chrisman, and Niemann, with support from the Barker Fund of the University of Wisconsin Foundation.

A geographic or land information system (GIS/LIS), as represented in figure 4.2, has an emphasis different from that of a multipurpose land records system or cadastre. The difference is in the degree to which the GIS/LIS systems emphasize the location of land features without discrimination between features, while land records systems emphasize the nature of land interests, or property rights, associated with parcels. The GIS/LIS systems strongly emphasize location aspects of land features and the provision of location-based services.

Many benefits result from the ability to locate, represent, and combine land feature location layers in the manner represented in figure 4.2. Provision of all kinds of location-based services in a more efficient and effective manner is a major benefit of that ability. The provision of these location-based services by GIS/LIS has a history of its own, distinct from that associated with the history of land records in support of land planning and management.[5]

However, the development of GIS/LIS hinders and supports location-based services. The parcel layer has become one layer among many in the concept represented in figure 4.2. Adherence to this concept has the effect of diminishing the importance of the parcel in the development of data and records systems that service land planning and management. This diminishment is not a problem as long as the major interest is in the location and extent of land features, including the surface boundaries of parcels. It is a problem when there is a need to combine data about the location of land features with records of land interests. The concept depicted in figure 4.2 is not conducive to the development of land records systems for management of land and its resources in twenty-first-century America.

It should be noted with great emphasis that the concept depicted in figure 4.2 makes no reference to title, assessment, or other documentation of land resources. This concept also makes no reference to the use of a unique PIN as a means for connecting the location of land features with the nature of land interests associated with these features. In this regard, the concept depicted in figure 4.1 is more relevant today.

The unique PIN is the practical basis for connecting records of *both* the nature and extent/location of land features and interests. It can be attached to documents that describe in words the substantive aspects of land parcel rights, restrictions, and responsibilities that apply to parcels.

There have been efforts to promote the widespread use of PINs as a means to link records and information about both the nature and extent of land features and interests. These efforts have been made even as GIS/LIS developed to serve location-based services (Vonderohe et al. 1991).

Recently, the major federal legislative response to the banking and lending activity associated with the recent economic downturn was the Dodd-Frank Wall Street Reform and Consumer Protection Act (Dodd-Frank Act 2010). A provision of the Dodd-Frank Act revised the Home Mortgage Disclosure Act of 1975 (HMDA). The HMDA requires certain institutions to collect, to report to federal agencies, and to disclose to the public data about applications, originations, and purchases of home mortgage loans. Among the data elements proposed for collection and reporting are the property value and a PIN. Modern technology makes this objective practical. It is not unreasonable to have mortgage documents presented for recordation in the register of deeds to have a PIN.

The parcel retains a uniquely important place in people's attitudes and actions regarding the use of land and its resource. The nature of rights, restrictions, and responsibilities give parcels a unique status that distinguishes them from all other land features in a modern land records system that serves planning and management. An emphasis on location-based services, with its emphasis on observable locations, directs attention away from the cultural importance of parcels and their associated land interests. This is particularly important at the local level of concern and government. This effect may explain, in part, the failure to overcome the institutional land records problems of the type identified in the 1980 and 2007 NRC reports.

## 4.7   Conclusion

American land records and information systems were created along with the processes established to make land- and resource-use determinations. These systems reflect and channel attitudes and values that Americans hold regarding land and its resources.

The several independent systems constitute the land records and information institution. The constituent systems include those of the local register of deeds and assessor's offices and other record-keeping public offices. Private processes are also part of the set of land records systems, including those of abstractors, title attorneys, surveyors, insurers, lenders, and brokers. These form an institution in the sense that they have developed to the point that they are a custom, practice, or behavioral pattern of importance in the lives of communities and societies.

American land records and information systems acquired important characteristics as they developed. These characteristics are important when the current systems are examined and when changes in the systems are proposed. They determine what practices are important and why, how they can and should be changed, what priorities should be established, what techniques are appropriate, what results are likely, and what the impacts might be. These characteristics also indicate changes that are both desirable and practical.

These characteristics are the following:

- Level of government. The major venue for land-use decisions in America is the local level of government—city, county, and regional. This venue is consistent with the American preference for private land decisions. When collective land-use decisions are recognized as needed, Americans prefer that the action be at the local level, closest to the individual citizen. The current strength of this preference is evidenced by the facts that Congress passed no national land-use legislation among the many environmental laws passed in the 1970s, that no such legislation has been considered since that time, and that the general expectation is that zoning be done at the local level. Each community has its own land concepts, values, and attitudes. Each strives to assert its distinctiveness and preserve its prerogatives. Whatever the merits of this American preference, it remains sufficiently powerful that it must be seriously considered in any proposed program of changes in land tenure and land records systems.

- Public land records and information offices are important local political institutions. Local assessors and registers of deeds are examples. Each community seeks to maintain the local government prerogatives regarding the local government land records systems. Again, whatever the merits of distributing land records among private professions, local government, and national government offices, the importance of local political institutions must be considered in a proposed program of change. When local political institutions are involved, land records system changes that are generated at the national level will face obstacles when implementation is sought at the local level.

- Three types of land records systems have emerged. These are title, assessment, and resource records. Each has its own set of personal, professional, administrative, and economic aspects:

  - **Title records**. American title records systems emphasize voluntary deposition of privately arranged property conveyances in a local document recording office, usually called the register of deeds. Officials give the document a cursory examination for the names of an apparent grantor and grantee, a property description or other reference to location such that there appears to be a chance to find the land on the ground, and the type of conveyance (deed, easement, mortgage, etc.). Acceptance by the register of deeds for recordation gives the document a legal status or authority. Recordation gives notice to the public of the document's existence and of its contents, but the official makes no effort to validate or assure the asserted names, description, or conveyance type. The location of all the documents for a parcel that have accumulated over time, a summary of their contents, the validity of the assertions in these documents, a summary of the meaning of these statements regarding the status of who has the right to decide about use of the land, and an assurance of this summary is done by private professionals. These professionals include title abstractors, attorneys, insurers, and surveyors. This process was well established and in common practice in America by the end of the nineteenth century. The basic process and the roles of public and private actors have not changed substantially since then. Computer technology has sustained the viability of this process. It merely speeds up a manual task instead of exploiting the technology to reengineer the system.

- **Assessment records.** Counties and cities are responsible for real property assessment. The authorized duty of the local property assessment official is to estimate the value of a property (market value is the preferred method; other methods are possible), apply the local property tax rate to that value, send a tax bill to the parcel's owner, and receive and process a check from that party. The owner could have a condominium, time-share, or other type of taxable interest. Activities that go much beyond this can be regarded as exceeding the authority granted to the official by the legislature to the agency in the statute. This administrative law restraint can have an effect on the extent of land-records system development in the assessor's office.

  Records are used to assist the assessor in the execution of authorized tasks. By the end of the nineteenth century, some assessors had begun to use maps to help them organize the growing array of assessment information, visualize the arrangement of parcels and other features, estimate parcel size, and help keep an inventory of the location of other, useful data. The assessors also used these maps to ensure that all land was taxed. Use of geospatial technology and digitization of all of the parcel maps led to the discovery of untaxed taxable properties. The typical map, then and now, presents the approximate boundaries of the property on the earth's surface. The map also provides the name and address of a party to whom a tax bill can be sent and contains other information useful to the assessor in the execution of the assessor's property tax duties. The local property tax map, often called a *parcel map*, has no legal status in a court dispute over the limit or boundary of the ownership rights for a parcel, it does not represent an authorized name of the owner, and it does not represent an authorized statement about the parcel boundaries. The map merely presents an approximation of both ownership and location of rights. This status has not changed since the end of the nineteenth century.

- **Resource records**. Resource records contain data and information about the cultural and physical features of land and its resources. Many natural resource features have ambulatory or hard-to-identify boundaries, sometimes both. Ambulatory features are those whose location changes slowly under the influence of both natural and human forces. Examples include rivers, wetlands, and flood hazard areas. Hard-to-identify features are represented by such things as soil types, earthquake zones, and radon zones. Wetlands, for example, are both ambulatory and hard to identify. Locating boundaries requires experts with professional training and experience.

  Location, description, and representation of these natural features present problems different from those associated with title and assessment records. First, identification of natural feature boundaries on the ground requires expertly trained professionals (botanists, soils scientists, surveyors, photogrammetrists, remote sensing scientists, cartographers, etc.) with skills different from those needed to describe and interpret title documents (lawyers, insurers, etc.) and establish property values (assessors, brokers). Second, concepts associated with these natural features have, in some cases, changed radically during the twentieth century and continue to evolve. Additionally, the boundaries of the natural resources whose locations are represented on the maps are often ambulatory, such as a coastline. The map represents conditions at a specific time.

  Third, and importantly, descriptions and maps of natural resource features generally do not acquire the same degree of legally recognized authority given to title and assessment records. For example, a wetlands map provides an approximate representation of a boundary location. The map provides advice to government officials and interested parties at a preliminary stage of a process where a permit is needed to change the land use. However, the map has not been granted the authority to be the definitive, legal basis for the permit decision. The map becomes a part of the land-use drama in that it can be challenged by presentation of evidence that attacks the map's quality. It is much easier to challenge the wetlands map that has not acquired a legal status for decision making than it is to challenge a flood hazard map that has been given legal authority by a local zoning ordinance that designates the map as the basis for decisions about which parcels are eligible for low-cost flood insurance. The prevailing rule for those maps without decision-making authority is the developer's clarion call, "Do not depend on government maps. With good science, the boundaries can be negotiated."

The processes and practices that are the bases of the American land records and information institution have not changed substantially since the beginning of the twentieth century. Significant scientific and technical developments have made it possible to sustain these systems. However, the same actors are executing the same basic tasks now as they did long ago. In this regard, modern American land records and information institutions exhibit a pattern of arrested development. They are subject to forces that prevent their motion, progress, and growth.

American land records and information institutions can be so described for a number of reasons:

- Stature-based administrative duties and the roles of public and private actors are essentially unchanged from those of long ago despite changes in the complexity of land- and resource-use determinations.

- It is difficult to assemble and combine land data from more than one independent land records institution in order to create the data and information for twenty-first-century land- and resource-use determinations.

- Special efforts beyond the routine duties of public and private parties are necessary to gather the desired land records for land-use determinations.

- Revolutionary changes in land measurement and information technology have not significantly altered the institutional practices and processes.

- Scientific and technical developments have emphasized the location aspect of land features at the expense of attention to the nature of land interests (rights, restrictions, and responsibilities).

- Private-sector actors dominate the activity of assembling land data and information from a variety of public offices despite the twenty-first-century description of land as "a commodity affected with a public interest" (Babcock and Feuer 1979).

Many of these conditions are cultural problems not easily susceptible to technological remedies. Solutions to these problems require changes in the way that people perceive land and its resources and how they channel their practices into land and land records institutions.

History and other cultural factors suggest the prime elements needed for modernization of the American land records and information institution:

- The division into title, assessment, and resource subsystems

- The division of labor between public and private actors

- The status of legal authority for land documents, records, and maps used to make land and resource determinations

- The role of land records management among the various levels of government

Unless these problems are addressed, appropriate land data and information will not be available for twenty-first century American land-use determinations. Production, assembly, and distribution of land records must be done efficiently. These activities must also be done effectively in the sense that the information package generated must meet the focused demand for land data and information as it is actually used in the processes for making land-use determinations. These activities must also be done equitably, meaning that these products should be open and accessible to the largest set of citizens ready and willing to contribute to and use them.

New land-use problems make it clear that modernization of American land records and information concepts, practices, processes, and institutions can no longer be postponed. These new problems include global warming, disaster management activities associated with hurricanes and tornadoes in urban areas, the need for new sources of energy and its means of delivery, challenges in land markets after the recent mortgage crisis, urban redevelopment challenges, and unmet pollution problems. These problems are part of changing activities that demand attention to efficiency, effectiveness, and equity in all aspects of American society. Land records and information system problems are an essential, often overlooked part of that society.

# Notes

1.  See note 2 in chapter 1.

2.  In 2003, the eastern Canadian province of New Brunswick began the extraordinary transition from a system of deeds recordation to title registration. The province assures the status of land interest ownership. This long-planned transition was associated with changes in the land records and information system. An overview of the new system by the provincial agency states, "Land Titles is a parcel based registration system which utilizes a Parcel Identifier (PID) to

uniquely identify parcels of land and to record interests in that land. Unlike the Registry System, which has existed in New Brunswick for over 200 years, once parcels have been converted to Land Titles as 'registered land,' the interests of individuals and enterprises in a parcel are guaranteed by the province" (New Brunswick 2012).

3. The Real Estate Settlements and Procedures Act of 1974 (RESPA 1974) called for a study by the US Department of Housing and Urban Development (HUD) of innovations that would reduce real estate transfer costs. A contract with the consulting company Booz Allen Hamilton resulted in an extensive study of the issue. It also included a study of the feasibility of a title registration system (e.g., a Torrens title system) in America in place of the existing title recordation system. The Booz Allen Hamilton report remained in draft form.

4. Modernization of land records has been a subject of considerable attention for many years (MOLDS 1975, 1979).

5. The development of geospatial technology and location-based data and information systems generated an early history of GIS/LIS (Foresman 1997). This history focused on the introduction and diffusion of computer-based geospatial technology. It did not focus on the long-standing problems of connecting the products of land records and information with the land management and planning processes. A land records world prior to the use of computers was not considered even though there was a considerable history of the connection between land records and land planning and management. A history of GIS/LIS in 1997 seemed premature and narrow if the objective was to place land records and information in the context of land planning and management in twenty-first-century America.

# Chapter 5
## Land governance in America

The nature and scope of governance in a society is a dominant factor in the development of a land records system. Governance encompasses the attitudes and practices that citizens adopt and use to organize and operate their governments. Governance extends to who acquires the authority to determine government activity regarding people, places, and actions. Other aspects include the preferred balance between individual and collective action, the process by which government authority is distributed, and who benefits from government activity. America, like any society, has a distinctive history of governance attitudes and practices.

Each community, at the national, state, regional, city, and village levels, has a distinctive and sometimes unique set of preferred attitudes and practices that are the bases for its political, social, and economic institutions. Some of these attitudes and practices appear in laws and legal process. Other attitudes are reflected in the commonly understood way that power is exercised. Even small towns can have widely divergent values, interests, and practices. In some small towns, well-off people on one side have a well-established order; and the poor on the other side have entirely different ways of doing things even when they are subject to or subjected to the power structure of the well off. Together, the preferred patterns of public and private attitudes and practices establish a pattern of citizen participation in many forms of governance in the community, including land planning and management. This pattern ultimately affects the land records institutions.

In the spectrum of governance types, three theories on the spectrum illustrate its breadth:

- the elitist theory

- the pluralistic theory

- the participatory theory

The elitist theory describes a government dominated by governing elite and their chosen officials. Citizens are involved when the governing elite perceive that citizens need to be given the information necessary to accomplish the tasks established by the government. This system favors the wealthy and powerful and those with professional expertise.

The pluralistic theory describes a system of governance where citizens are involved in choosing those who exercise power—those who make statutes, rules, judgments, and so on. Citizen power is found in the power of groups. Political parties are the major example of group power. Citizen and other interest groups often are important actors.

The participatory theory describes a governance system where the individual and the small group can participate significantly, directly, and effectively in government plans, decisions, and actions.[1]

The United States is sometimes described as a pluralistic society (Smith 2009). Others assert that the elitism theory better describes policy formation and implementation in the United States (Mill 1993).

"A characteristic of American political culture is that people value the opportunity to take part in the decisions that affect them. Add to this the traditional American skepticism toward technical experts and centralized power and the challenge to policy makers is clear" (Fiorino 1995, 6).

Whatever description is most appropriate, it appears that many Americans aspire to a governance pattern that enhances citizen participation in the many community decisions that determine the conditions of their lives.

American history reveals that the expectation of citizen participation in decisions is particularly strong regarding land planning and management. The strength of this expectation is revealed in the preferred pattern of local, site-specific land- and resource-use determinations and in the local land records and information institutions that are a part of the government infrastructure.

## 5.1    Origins and sustenance of American land governance attitudes and practices

The attitudes and practices regarding citizen participation in governance have a long history. They began with the earliest English settlers in North America. The settlement of Massachusetts Bay in 1630 by the Puritans was accompanied by a revolutionary change in the pattern of governance among Englishmen (Morgan 2006). Within months of initial settlement, the leaders who had acquired a royal charter and held all the power willingly chose to expand membership in the group with political rights. Their view was that many should share in the choice of those who would make the laws and select the governor. All freemen, most of whom had no political rights under the charter, were allowed to choose a governing group by vote of all. They converted the governing group and its director into a legislative assembly that chose the governor (120).

Why did the leaders allow this expansion, and what is its purpose in designing a modern land records system? One suggested answer is that community members in 1630 were committed to a covenant where salvation depended on each individual's commitment to be bound by God's laws. Another possibility is that there was a need to have a government that ensured the covenant was enforced because the people "did not have enough virtue to carry out their agreement without the compulsive force of government. They must decide among themselves what form of government they wanted and then create it by a voluntary joint compact" (123). The leaders were not so committed to popular rule such that they believed that the people were entitled to determine the form of government. The leaders did believe that "people would submit to the leader if they had a voice in choosing him, especially a Puritan people well educated by their ministers in the principle of a government based on covenant" (126). "There was a danger, of course, that the people would choose the wrong kind of man to rule them" (126). Nevertheless, there was a confidence that the ministers would give the people sound advice and instruct them about the kind of men who were best fitted to rule. There was reason for concern about the danger. The "three-thousand mile moat on the one hand and boundless free land on the other offered strong temptation to kick over the traces and defy every kind of authority" (127).

The Pilgrims at Plymouth had earlier adopted a similar, open participatory governance, beginning with the Mayflower Compact in 1620 (Philbrick 2006). The Pilgrims' chain of authority from the king was not as clear as that of the Puritans. The Puritans were composed of a mix of believers and nonbelievers, and there existed a greater need for an agreed-on authority. The Pilgrims' leaders had less power to relinquish. The Pilgrim and Puritan experiments with participatory governance occurred after the earlier Jamestown settlement. Initially, the Jamestown settlement of 1607 was not accompanied by a strong commitment to participation in governance by all its members. The Plymouth and Puritan settlements quickly moved from collective to private land ownership and management while retaining a commitment to participatory governance.

The community concept of a common religious covenant in the seventeenth century has a twenty-first-century analog in the form of a common social contract or dominant social paradigm. The attitudes and practices that generally characterized early American governance and land planning and management specifically have persisted for several centuries. The development of private property concepts, the widespread distribution of privately held land under the PLSS, the development of local governments and schools, and the emphasis on local government land-use controls all suggest continued American aspiration for active, direct citizen participation in land governance. Active, direct participation does not have to mean voting in town meetings. This participation can take the form of citizens who are well informed by the products of geospatial technology and who have laws and legal processes that ensure that governments actively receive and incorporate informed citizen input into government actions.

The future of land and resource use and of land records and information systems depends on the system of governance that will prevail in the twenty-first century. Several questions arise, including the following:

- Does the future lie more with elitism, pluralism, or participation?

- What role will land records play in the future?

- Will land records systems reflect the governance pattern?

- Can design of land records systems influence or just reflect governance?

# 5.2    Public administration

Public administration is about how plans, decisions, and actions are made and implemented by legislatures, agencies, and courts. The administrative agencies usually are charged with the day-to-day details of governance (Fox 1986, 1).

Agencies often are delegated considerable power and discretion by a legislature's enabling statute. This often is by design of the legislature that relies on the technical expertise of the agency to establish the detailed rules within the structure of the statute's broad language. The ability of the US Congress to delegate power to an administrative agency was established in the US Constitution in the so-called "necessary and proper clause" (US Constitution, Art I, §3, §§18).

The following are important agency activities:

- Make policy—the means societies use to decide how resources will be used and who will benefit from their use—in the form of administrative rules and regulations within the confines and restrictions of an enabling statute. Frequently, statutes are written in broad language; therefore, agencies often have a considerable degree of flexibility in their choice of the details that make specific the broadly stated words of the statute.

- Examine, decide, and implement applications for benefits and permits established by the statutes and the administrative rules.

"For the better part of the twentieth century, the public bureaucracy has been the locus of public policy formation and the major determinant of where this country is going" (Henry 1975, 3). The study of public administration encompasses theory and practice "designed to promote a superior understanding of government and its relation with the society it governs as well as to encourage public practices more responsive to social needs and institute managerial practices on the part of the public bureaucracies that are substantially attuned to effectiveness, efficiency, and, increasingly, the deeper human requisites of citizenry" (4).

## 5.3   Land administration

Land administration is defined in the United Nations Economic Commission for Europe (UNECE) Land Administration Guidelines as, "The process of recording and disseminating information about ownership, value, and use of land when implementing land management policies" (Williamson et al. 2010). This definition has the considerable virtue that it describes a connection between land records in the context of land planning and management.

Government infrastructure is more than the roads, bridges, power lines, and other tangible items typically associated with development. Also included are governmental processes, practices, and institutions that regulate public and private land actions and the land records and information systems that support these actions. Land administration is part of a government infrastructure that extends to all the activities necessary to implement land-related policies and land management strategies (Williamson et al. 2010).

Public land planning and management in America, including the land administration component, have distinctive characteristics. Speaking of the environmental statutes that are a major part of land administration and their administration, one observer has noted:

> Much environmental law today is "fine print" in the pejorative sense.
>
> Students must be made to confront the likelihood that their initial understanding of each environmental control scheme is misleading, because the scheme will be shown to be vastly different once the fine print has been explored.

"The devil is in the details."

For every statutory section, the administrative rules and regulations are at least an order of magnitude more complex than the statute itself.

Beneath the rule lie numerous interpretations, caveats, exceptions, guidance documents, regulatory preambles, agency manuals, letter rulings, policies, precedents, and manifold other administrative utterances.

The generality of environmental law practice today takes place deep down in the interpretative pyramid, not in the statutes and regulations at its apex.

The issue is no longer what the words of the statute seem to say, but what the agency has said that they say, and whether that administrative construction is within the broad parameters of agency discretion.

Many, if not most, complex distinctions in environmental rules are the fossilized evidence of a past political deal.[2] (Stensvaag 1999)

These observations suggest the nature and extent of the complexity associated with land planning and land administration. They also point to the problems that must be considered in efforts to modernize land records in ways that promote citizen participation in land planning and management.

## 5.4  The importance of local government and local land records in land governance

A major expression of American governance attitude and practice is the historical American preference for private determination of land and resource use. This has several implications. One is continuing emphasis on land interests (property rights, restrictions, and responsibilities) associated with each parcel. As a result, each site, parcel, and area is regarded as unique in real property law. Therefore, land records and information used for land planning and management must be legally appropriate at the parcel level and reveal both the physical conditions and the status of land interests. A second implication is the result of the significant twentieth-century development of community involvement in land planning and management in the form of public land-use controls and environmental laws. The preferred venue for most land-use control and site-specific environmental legislation was and remains the local level of government in America.

It is not surprising that Americans who prefer private determinations of land and resource use should also prefer that public land-use determinations be made by a government as close as possible to those individuals and groups who are most directly involved or affected by the determinations.

In the 1970s, when most of the major federal environmental laws were passed, attempts to establish national land-use control legislation were unsuccessful. States and local governments remained the primary venues for public land-use determinations. More typical of federal statutes that influence, but do not control, site-specific land and resource uses are provisions of the Clean Air Act (CAA). The Act imposes federal limits on the allowed concentrations in the ambient air for a set of ubiquitous, harmful pollutants from many sources. However, the Act also leaves to the states the power to determine how these ambient air quality standards are to be achieved. This means that the states, by common deference to local governments, establish how parcel- and site-specific land uses are determined.

This shared authority between the federal and state governments is a characteristic of federal environmental statutes. It is a reflection of the ongoing debate over the distribution of federal and state power, or *federalism*. This shared authority reflects the strongly held attitudes that land and resource use at the parcel level should be determined locally and not at the federal level.

# 5.5   Administrative procedures and land governance

Active citizen participation in governance is promoted by the federal Administrative Procedures Act (APA) and its state equivalents (APA 1946, as amended). The federal law requires a public notice and comment process before agencies take final action on a plan, a new or altered administrative rule, or an application for a permit. Notice of proposed agency action must appear when the plan, rule, or permit action is first under consideration, well before final action is taken. All interested parties are to have access to the proposal and the supporting data and information. The parties then have a period during which they can examine the proposal and the supporting data, identify data and other errors in the proposal, and submit comments and alternative plans and actions to the agency. Discussions between agency officials, a permit or plan applicant, and affected parties may occur prior to the notice and comment process. These discussions do not overcome the legal requirement for the public notice and comment process.

The notice and comment process is ubiquitous in American government. It applies to many administrative actions that affect property rights. The notice can be a mere posting in a local newspaper or a widespread distribution on an agency web page. It can cover activities such as quiet title action, a proposed land-use regulation, an application for a permit to build an industrial facility, or a zoning change. The governance challenge is to assure that all affected parties are actively notified at the beginning of the process instead of leaving them without notice or requiring them to devote time and effort to a continuing search for notices. Geospatial technology makes it possible to identify all interested or affected parties and actively notify them.

All states have some form of an administrative procedures act for their state and local governments. The state requirement for a notice and comment process usually extends to all levels of government and for many government agency plans, rules, and permit decisions. This process expands the reach of the act and its provision for citizen participation far down into the administrative process in government agencies.

The public notice and comment process has been described as having disadvantages as a mechanism for citizen participation in governance. The public hearings may be loosely structured, where citizens hear agency proposals and respond reasonably and unreasonably. Preliminary discussions among parties can occur before the process, thereby narrowing the options. Hearings and comments can be late in the process. Participation is often strongest among opponents (Fiorino 1995, 96).

Public hearings are only one part of the notice and comment process. There are other opportunities for citizens to acquire and submit land records and information relevant to the agency administrative action. These opportunities occur when the agency posts all proposed rules or permit applications or other proposed actions on its web page as part of its process. The agency then posts all received comments, thereby making all comments available to all interested parties as the process continues. The agency cannot easily ignore these comments if they address significant issues backed by appropriate records and information and their substantive analysis. This process is more than the open, public hearing that can result in assertions of public opinion that may not be respectful, dignified, and professional.

Congress passed the National Environmental Policy Act (NEPA) in 1969. This act requires a report, commonly called an environment impact statement (EIS), that each federal agency must prepare to accompany each "major federal action significantly affecting the quality of the human environment" (NEPA, §102). Each state chooses to have or not have a state equivalent for its state agencies. However, in many states without a state equivalent to NEPA, there are specific state land and resource statutes that require the development of specific data and records before plans and decisions are made and actions are taken by the agency with delegated power. The requirements of these federal and state statutes, and the subsequent regulations, constitute a major demand for land data and records in land planning and management. The data and records are now commonly expected by all to appear for public scrutiny before final agency action.

An agency web page used in this way encourages and provides opportunities for all citizens to obtain records and information and to communicate with the agency and with one another in a venue common to all. Geospatial technology can be used by any interested party to analyze records and information and introduce the results of a sophisticated analysis into the process, thereby providing an alternative to the noisy, public hearing part of the notice and comment process.

A legislature or council can use a web page for notice to citizens of proposed legislative branch actions in a manner analogous to the notice given by agencies of proposed agency administrative actions. The legislature or council can give notice to the public of proposed legislation or other legislative activities and provide citizens with a venue for response and communication prior to creating the legislation. At the least, the legislature can provide notice of proposed legislation. At the most, the process can provide for responsive comment by citizens. A good way to begin this development is to use geospatial technology to give notice to citizens of proposed legislative actions that affect the nature and extent of land interests.

A land records system designed specifically to serve all parties in a notice and comment process will improve citizen participation in land planning and management.

## 5.6   Open government, open records, and participatory governance

A major aspect of American governance is that citizens expect to know what their governments are doing. The historian Henry Steele Commager wrote that "the generation that made the nation thought secrecy in government one of the instruments of Old World tyranny and committed itself to the principle that a democracy cannot function unless the people are permitted to know what their government is up to" (Commager 1972, quoted in EPA v. Mink 1973).

Open government and participatory governance require citizens to have the practical ability to acquire the data, information, and knowledge used by government officials to support and explain their actions in the normal course of implementing their authorized duties. Officials may want to provide citizens with material chosen by the officials and in a form chosen by the officials. The choice by government officials of which data and information to provide is consistent with an elitist form of governance. When challenged for more information, officials may refuse with the complaint that citizens are not grateful for that which they have received. This response is inappropriate in a society where citizens expect to know what their governments are doing.

Modern geospatial technology makes it convenient and practical for an agency to create a website that presents a large array of government records and information. Agency managers and planners can use these websites to present material that they believe to be most appropriate in substance and form for use by citizens in planning and management. Citizens receive that which the officials think is appropriate for the citizen. The same geospatial technology also makes it convenient and practical for an agency to deliver to citizens all the open records and information actually used by the officials to execute their duties.

Americans do not readily defer to the judgments and actions of public officials. This behavior has a long history regarding land planning and management as represented by the preference for privately arranged site-specific land-use determinations and resistance to national government land-use controls in favor of state and local controls. Continued choice by officials of what material to place on government web pages cannot easily overcome long-established American preferences and may contribute to extended or growing citizen hostility to officials.

A major legal mechanism for knowing what a government is doing is a freedom of information, or open records, law. Freedom of information and open records acts make those data and information used by governments in the normal execution of their authorized duties conveniently available to all citizens. The pattern of these laws is that data and records, in the form used by the government, are to be provided to citizens in an unfettered manner.

Security and privacy are provided for in these laws. Specific exceptions to openness appear in the statutes. Courts are the final arbiters of the balance between openness and exceptions to openness. Openness is given favor and emphasis when courts interpret openness broadly, while interpreting exceptions specifically and narrowly.

The major federal statute is the Freedom of Information Act (FOIA) (1966). State and local governments have analogous statutes for their agencies, often called open records laws. The US Supreme Court interpreted FOIA to be designed to allow citizens to obtain records and information from the hands of officials who are sometimes unwilling to provide them (EPA v. Mink 1973).

Freedom of information and open records laws make it possible for interested citizens and groups to more actively participate in public land planning and management. Citizens have access to large and significant sources of records and information. Widespread access to powerful geospatial technology can be used for sophisticated analysis and representation of the computer-stored land records and information. These results can be introduced in the land planning and management process. The notice and comment process provides a common venue for doing this.

Open, unfettered access by citizens to land records and information used by governments in land planning and management is consistent with traditional American concepts of both property and governance. Open, unfettered access enables citizens, groups, organizations, and officials to know what can be done with land resources, what the impact will be, and who has the power to decide. Open, unfettered access also makes it possible for creative people to take the material and make new, valuable uses of records and information by creating new products, services, and institutions. With this open access, land records systems can play a role in the development of participatory government, especially at the local level of government. This can be a model for citizen participation at all levels of government. Increased citizen participation in governance can promote positive attitudes and actions regarding governments and governance.

# 5.7   Conclusion

It is now common to see an agency website with a proposed agency action, a full set of material submitted in support of the action, and all comments during the course of a required notice and comment process. Freedom of information and open records laws do not give an agency the discretion to withhold any of this material unless it is specifically exempted by statute. Posting all this material on a website is an improvement over a process wherein only a public hearing is conducted. This development provides an expanded venue for full citizen participation in governance, and it expands the nature of the venue to one that goes beyond the often noisy, chaotic public hearing where citizens can speak in a variety of styles while officials politely listen and proceed with their work.

Unfettered access to the land records and information assembled by agencies in the normal course of all their authorized duties means that the materials are available not only to the wealthy and powerful but also to those who are poor or powerless. The high level of geospatial technology development, the large and increasing number of people who can use the technology, and the existence of interest groups willing and able to assist the poor and powerless means that unprecedented numbers of individuals and groups can take advantage of these developments to enhance their ability to participate directly in governance. This participation is particularly important for traditionally disadvantaged groups and individuals who can now work with those willing and able to help them begin to use the technology.

However, more is needed than access to records and information, even unfettered access, for full citizen participation in planning and management. Full citizen participation depends on several factors, including the following:

- Provision by agencies of government records and information used in the normal course of agency activity.

- Provision by citizens of records and information. Citizen volunteered records and information and their wide distribution is an increasing part of the modern, technological world.

- Establishment by agencies of formal processes to not only receive citizen-generated records and information but also to actively examine and give status to those records found to be appropriate for agency land planning and management activity. This active response effort provides a sorting mechanism that gives planning and management priority to specific data in a world overflowing with records and information. These processes are not well or widely developed.

Modern geospatial and information technologies make all these activities and the expanded venue for citizen participation in agency activity practical and common. Administrative procedures acts require citizen notice and comment prior to agency final action. Freedom of information and open records laws require convenient access by all participants to the data, records, and comments during the course of the process and to records and information held by agencies during the normal course of other agency activities.

The confluence of geospatial technology development, its widespread distribution and use, American administrative procedures, open government and open records laws, and the American system of environmental assessments prior to land-use change activity afford an opportunity to promote citizen participatory governance in land planning and management in twenty-first-century America.

The histories of land governance and the related histories of land and of land records provide a context for consideration of changes in the American system of land records and information. These histories remind us of the importance of people-oriented processes and the importance of technical developments. These histories direct attention to the idea that land records and information systems are not just an object but are also a structure for positive change. A modern land records and information system is both an object and a process. Consideration of changes in existing systems must consider both aspects.

Chapter 6 presents ideas about how to modernize American land records systems. The ideas consider the nature and scope of the representative problems discussed in chapter 2. They also consider the incentives for and barriers to changes in processes associated with the preferred American attitudes and practices discussed in chapters 3, 4, and 5. An overriding objective in the design of a modern system described in chapter 6 is a system with the major function of satisfying the demand for land records and information sought by all parties in the normal course of land planning and management.

# Notes

1. These three points represent a broad continuum of governance types. The descriptions of the points illustrate the way that many Americans perceive the development of governance since the beginning of the colonial period. American history and current attitudes and practices suggest a continuing desire for direct citizen participation in governance. This desire continues, however strong may be the perception that governance in practice is something else. The desire for direct citizen participation in land planning and management, especially at the local level of government, combined with geospatial technology can be used as a force for development of a modern American land records system.

2. The development of modern environmentalism and of modern environmental law and legal process has modified nineteenth-century attitudes and practices regarding use of land and its resources. We have chosen to emphasize the development of the traditional attitudes and practices in our discussions in this book. As the quotes suggest, existing law and legal process provide ample opportunity for the use of traditional attitudes in current land planning and management (Stensvaag 1999).

# Part III    Solutions, actions, and prospects

Part III describes a land records and information institution consistent with long-established, preferred American attitudes and practices regarding use of land and therefore appropriate for twenty-first century American land planning and management. We call this institution a modern, multipurpose American land records system (ALRS). This system is designed to overcome the institutional barriers to modernization of the chaotic American land records and information systems. An ALRS is distinctive in its emphasis on connecting records and information about both the nature and extent of land features and interests.

The major elements and attributes of an ALRS are not limited to those associated with location-oriented data. They also include the people-oriented processes wherein the records and information are gathered, managed, distributed, and, most importantly, used. The distinctive nature and scope of these elements in an ALRS are derived from the preferred American attitudes and practices regarding land and land governance.

The ingredients and attributes of an ALRS are defined and described in chapter 6. The economic, social, and political prospects for development of an ALRS are considered in chapter 7. Specific actions designed to develop and sustain an ALRS, consistent with incentives and barriers revealed by the histories of land, land records, and land governance in America, are described in chapter 8.

# Chapter 6

## An American land records system (ALRS)

Modernizing American land records and information institutions for twenty-first century land planning and management requires attention to both people-oriented, dynamic processes and the attributes of technology-driven products.

Citizens, groups, organizations, and officials are involved in processes wherein plans, decisions, and actions determine how American land and its resources are used. Geospatial technology generates a huge array of products that can be useful to all of the interested parties. Not all products are appropriate for determination of land use. The processes wherein land-use determinations are made distinguish and choose among the available data and information products in ways that are based on cultural factors rather than scientific factors. The result is that the operating land records and information system that has developed over many years does not adequately serve the needs of twenty-first century American land planning and management. Changes in both process and product are required.

Operating systems have successfully served many of the past and some of the current needs. Many individuals and organizations depend on the existing arrangements. Change is difficult in this context. Nevertheless, social, political, and economic forces external to the world of land records and information systems are imposing change on that world. These external forces are exhibited in the nature, pace, and scope of change in land and resource use. These forces are reflected in land planning and management processes. Changes in processes need to be coordinated with the attributes of records and information provided by the operating land records systems. Coordinating a demand for a preferred process with the supply of products is a challenge. It is a particularly difficult activity when modern geospatial technology has few limits on the nature and character of the products that can be supplied. However, in a world of limited resources, operating land records and information systems strive to coordinate and direct their efforts to efficiently, effectively, and equitably assemble and distribute the material according to the substantive and procedural planning and management demands. Building a modern records and information system requires attention to both processes and substance.

This challenge can be met by using the external forces to direct substantive and procedural change within each of the existing individual land records and information institutions. Coordination among the several institutions depends on constant attention to the priority of changes that are consistent with and directed toward the objective of better twenty-first century land planning and management.

The need for an improved system was revealed by a comparison of the recommendations in reports on land records systems sponsored by the National Research Council (NRC) in 1980 and 2007 (NRC 1980, 2007). Both reports recommended the establishment of nationally integrated land parcel data and maps. The failure to achieve such a basic and necessary objective after nearly three decades of often spectacular geospatial technology development indicates that we need to address people-oriented, institutional, and process aspects of land records and information systems.

A multipurpose American land records system (ALRS) is an institution designed, implemented, and sustained to provide connected land records and information about both the nature and extent of land features and interests. This connected material has attributes needed to support a multitude of land planning and management activities in a community. The processes and products of an ALRS conform to the preferred attitudes and practices of community members regarding land use and land governance.

An ALRS is distinguishable from other geospatial data systems (e.g., geographic, land, resource, and environmental information systems) that emphasize data, records, and maps of the location and extent of land features and land improvements. An ALRS differs from other spatial information systems in its balanced attention to records and information about both the location and property interests in land features.

Those who seek to modernize American land records and information systems in order to meet the needs for twenty-first century American land planning and management confront several challenges and tasks:

- Can changes be introduced that move the system toward meeting the needs of all citizens within the long-established American attitudes and practices regarding land, land records, and land governance?

- Is it possible to combine the *demand* for land records and information coming from citizens, groups, organizations, and officials who are active in land planning and management with the *supply* of location-based services coming from geospatial technology?

The demand for land records and information arises in the normal course of the law and legal process for land planning and management. This demand is a basis for a directed effort that sorts through the chaos of products that technology provides. It brings an order to the efforts of public and private individuals, groups, organizations, and officials who assemble and use land records and information to determine land and resource use.

## 6.1   Perspectives on an ALRS

Title records documenting the results of private negotiations over the transfer of land interests are part of an American land records institution. An *institution* is commonly defined as "a custom, practice, relationship, or behavioral pattern of importance in the life of a community or society." It can be "an ever present feature" or "an established organization."[1] Public and private actors operate this system in a long-established, public-private partnership. Records of publicly established land-use controls and other public actions are scattered among many agencies, hardly connected or coordinated, and often difficult or impossible to identify or locate.

Geospatial and information technology has created extraordinary new techniques and services for obtaining, managing, and distributing data and information about the location of land features. However, it has not yet been fully deployed to solve the problem of identifying, locating, managing, and distributing these land records and information that document the nature of land interests. The problem is rooted in the difficult task of connecting records and information about the location of land features with records of entities that have the right to decide how these features are used.

Existing land records offices—such as the register of deeds, assessor's offices, and engineer's offices—often operate as independent institutions. Each agency has a degree of freedom in its enabling legislation to establish operating rules and procedures. Communities are not in the habit of passing legislation that gives an office the power to command overriding cooperation among long-established, independent land records institutions. Cooperation is more often voluntary, as between a register of deeds and an assessor's office, and this cooperation rarely overrides traditional institutional practice. Similarly, informal arrangements relate the activities of the public register of deeds and private title organizations.

Informal relationships between independent actors can result in uncoordinated, unconnected sets of land records and information in the community. The task is to bring together the generally independent institutions so that the needs for land records and information in twenty-first century land planning and management are satisfied.

A new institutional perspective is needed that rejects thinking of the land records systems as a set of independent land records institutions. Cooperation among institutions is driven by external forces in the form of a demand by all citizens for enabling data and information for use in community land planning and management. This approach is described as "thinking about versus thinking within institutions" (Heclo 2008, 4). This approach asks those who manage or direct a land records institution to focus on the major objective of satisfying the records and information demands of all those involved in land planning and management in the community. A narrow focus on providing the types of traditional products and services within traditional, existing land records institutions represents thinking within an institution.

To think institutionally about a land records institution rather than within a particular land information institution is to constantly ask the following:

- What are the preferred objectives for use of land and its resources in the community?

- How do the community land records and information support all those interested in defining and achieving these objectives?

Thinking about a land records and information institution in the context of twenty-first century land planning and management requires attention to existing long-established American attitudes and actions regarding how land and its resources are used and who determines that use. Americans prefer that individuals make land-use decisions without government involvement and that collective decisions be made at the local level of government, nearest to the individuals.[2]

This long-established condition means that knowledge of the nature and extent of land rights, restrictions, and responsibilities—the land interests, or property rights—is as important as knowledge of the location and extent of these interests and land features. An ALRS must reflect this knowledge.

Design and implementation of a focused, connected land records institution also requires constant attention to the many and often changing uses for land and its resources. A modernized land records institution must have processes that actively respond to documentation of new and changing land uses. For example, reports of and community response to invasive species requires both reports of observations and a public process that validates the observations as recognized bases for new, collective actions that can or will affect many people and parcels. It is not enough to have a repository of observations.

Land-use changes are often the result of administrative activities associated with public land planning and management. Land-use control legislation is an example. Records of these activities become the bases for organizing the records and institutions in a modernized American land records system because these public actions significantly affect all land-use determinations. A modernized land records system has procedures that actively identify and capture these changes and the documents that fully describe them.

A major objective of a modernized American land records and information institution is to bring together *both* the nature and extent of interests in land and its resources when, where, and how that information is needed for citizen participation in land planning and management. This is a focus on the demand for and not just the supply of land data and information.

# 6.2   A caveat about modern land records systems and multipurpose cadastres

A *cadastre* is a "record of interests in land, encompassing both the nature and extent of these interests" (NRC 1980). This succinct definition captures the important balance between records and information about the location of land features and records about the allocation of decision-making power regarding use of the features. A cadastre is concerned with records of the status of land interests and with data about the location of other aspects of land, such as geography, environment, value, culture, and natural resources. A cadastre is distinguished from other geospatial information systems by the degree of its emphasis on land rights, restrictions, and responsibilities. A cadastre is not just a map of parcel boundaries.

Cadastres have a long history in other countries. Many countries give special attention to records of public and private land interests (title cadastres). Others countries add land values (fiscal cadastres) or add information about land resources. In many countries, cadastres are part of a system that provides government recognition and assurance of the status of parcel ownership, parcel boundaries, and parcel values. Use of the term *cadastre* in the United States is uncommon, confined to a few agencies or professions, and inconsistent or misleading.[3]

Government recognition and assurance of the status of title and boundary are not a common feature of the American system of land records.[4] Prospects for increased, widespread government assurance of the status of title and boundary in the United States seem unlikely at this time. However, modernization of American land records and information systems can and should be designed and implemented in ways that do not exclude this option for the future.

Periodically over the past thirty years, *cadastre,* or *multipurpose cadastre,* has been suggested as a term for a modernized land records system in the United States. However, the term has not captured the attention of a broad American audience, has acquired a negative aura, or is used to mean something narrower than what has been defined here. Significantly, many knowledgeable Americans use or understand *cadastre* narrowly to mean a parcel map or parcel fabric.

The term could be used to mean a modern, multipurpose land records and information system. While striving to retain the possibility that *cadastre* will obtain favor in America, those involved in the system choose to use the terms *multipurpose cadastre* and a modern land records systems that we call an *American land records system (ALRS)* interchangeably.

# 6.3    Ingredients of a modern land records system

A modern land records system, or cadastre, serves the demand for land records and information in twenty-first century land planning and management. The system design and its implementation consider two issues. One is what records should be included (ingredients, discussed here). The other is what characteristics the records should have (attributes, discussed in section 6.4).

Answers to these questions are based on the demands for land records and information in the normal course of land planning and management in a community. The design should also recognize that the development of geospatial technology has not, by itself, solved many of the institutional problems that inhibit the assembly of appropriate data needed at the time and place needed for land plans, decisions, and actions.

The ingredients in a modern land records and information system, or cadastre, should include the following:

1. **Title records.** These are the documents that record title and other land interests that are the result of privately arranged land rights, restrictions, and responsibilities.

   The title records institution in America contains the public repository for those documents prepared by parties in a transfer of privately held land interests.

The land interests are those depicted in the bundle of rights model shown in figure 3.1. A government can be one of these parties when it exercises its sovereign role as a landowner. The government repository of these documents (and other documents such as mechanics' liens that apply to the status of private ownership of land) in America is the register of deeds office at the county or town level of government. The office is a repository for documents that parties choose to deposit there with the intent to give notice to the community of the document's existence.

A long-established system exists for locating, abstracting, and evaluating the documents that have accumulated over a long period of time in the repository and apply to a particular parcel. The register of deeds office is designed to provide a public place where these documents can be deposited and made available for public scrutiny. Subsequently, these documents can be located and evaluated in an effort to indicate who owns the many land interests in the privately held bundle. The office provides no assurance of the validity of the document or of the ownership status of the land interests.

The public register of deeds office exists in a symbiotic relationship with the private-sector actors in the effort to determine and express the status of land ownership. The private actors include the abstractor, title attorney, and title insurer. After locating and evaluating a set of documents in the register, a private party (the title attorney or title insurer) provides the buyer of land with an opinion of the ownership or distribution of ownership of land interests. Assurance of that opinion does not come from the government. Assurance of the opinion of title status comes from private title insurance or the private attorney's malpractice insurance, not from the government.

This title system's pattern of documents, actors, actions, and process were established in America by the end of the nineteenth century. The pattern of activity and the symbiotic relationships remain largely unchanged since the nineteenth century. Documentation of public actions, in the form of land-use controls and permit decisions that affect the status of privately held land interests generally are not included in the title process. The absence of these documents presents a major challenge to modernization of land records in America.

The title records ingredient in the existing land records institution has well-developed legal regimes, professional actors, practices, advocates, and critics. The issue for development of a modern land records system is how the existing title records system can be better connected to or integrated with the other land records ingredients, discussed next, in an effort to meet the demands for land data and information in twenty-first century American land planning and management.

2. **Assessment (fiscal) records.** These are the data and records that contain data and information used to determine the value of land and its improvements.

   Assessment of the value of parcels (the land, resources, and improvements) is done by both public and private organizations. Public assessment records in the American land records institution are those data and information assembled by the local government assessor's office, called the auditor's office in some jurisdictions. The legislated duty of the assessor is to estimate the value of a parcel for the purpose of a local property tax. A variety of data and information is needed for this purpose. This includes data about the physical status of land features, the distribution of privately held land interests, and the status of publicly established land-use controls. A parcel map is prepared as an aid in the efficient and effective administration of the office.

   It is now common for the assessor's office at the local level of government to be the primary source for a map of parcel surface boundaries and for a wide variety of other land features and the underlying data. The set of features varies widely according to the needs and resources of the local jurisdiction. The assembly by the local property tax official of parcel boundary maps and maps of other land features is what is commonly referred to as the geographic or land information system (GIS/LIS) in a community and commonly represented as in figure 4.2.

   The most important aspect of this system for the design of a modern American land records system is that these maps rarely have a legal status as the basis for a specific land planning and management action. A map from the assessor's files may be useful. Public and private individuals often use these maps for land plans, decisions, and actions. However, lack of a legally recognized authority designating the map as the basis for a plan, decision, or action means that data, information, and knowledge obtained from the map is not determinative in the land planning and management process. A party in the ad hoc, three-actor land-use drama can challenge the use of the data or map for the planning and management decision or action based on the right to present better data or maps. This gives considerable power in land governance to those with wealth or ability to attack the data obtained from the local GIS/LIS. It enhances the power of lawyers who represent the wealthy and powerful at the expense of those who collect or otherwise generate data and maps used by agencies.

3. **Natural resource records.** These are the data and information about the location and characteristics of natural land features (rivers, earthquake zones, wetlands, flood hazard areas, etc.).

4. **Infrastructure and other cultural records.** These are data and information about the location of manufactured or cultural land features (roads, telephone poles, building footprints, historical sites, etc.).

The data and information in categories 3 and 4 are most often found in geographic, land, resource, or environmental information systems. These systems and their data are characterized by an emphasis on the location of a variety of physical (e.g., soil, wetland, topography, parcel boundary, rivers, scenic views, forests, and habitat) and cultural features (e.g., surface parcel boundaries, buildings, power transmission lines, telephone poles). These are data and records systems for which there has been the greatest development over the last several decades. They are the products of the geospatial revolution. To the extent that these systems emphasize the location of features and interests, they tend to deemphasize the importance of public and private land rights, restrictions, and responsibilities.

5. **Administrative records.** These are the documents and records that describe the nature and extent of public actions affecting land interests.

These documents and records describe the nature and extent of a government's rights to control the use of land and its resources for a public purpose, to take land for a public use, to establish a property tax system, and to adjudicate disputes between private parties. These records often contain documents limiting land use within a prescribed distance from a natural feature, such as a shoreline. Often, the exercise of government power to control the use of land does not give adequate, detailed attention to the location of the regulated land use (e.g., "no buildings within 100 feet of the shoreline"). The plans, decisions, and actions alter the nature or allocation of existing land interests. These important records are distributed among a large number of government agencies with the legislated authority to exercise these government powers. These are the public actions and records for which it is most difficult to ascertain the existence, extent, location, and detail when an individual, organization, or agency seeks to determine the complete set of land interests for a parcel that encompass both public and private interests. These are the records for which there is the greatest degree of arrested development in American land records institutions. These are, nevertheless, among the important records that need to be organized and connected to other types of land records as population and other forces alter the uses of land and its resources during the twenty-first century.

The choice of this set of ingredients in a modernized, multipurpose ALRS is based on both the nature and extent of the data and records systems, the nature and extent of land concepts and records systems that have developed in America, and the desire for citizen participation in land governance.

The several types of data and information that should constitute a land records and information system should also have certain attributes, or characteristics, which are discussed in the next section. The data and information should be accessible, accurate, affordable, appropriate, authoritative, complete, consistent, coordinated/connected, institutional, uncompromised, interoperable, unique, open, precise, reliable, sustainable, timely, and verifiable.

Many of these attributes, and perhaps others, are the subject of considerable interest in the development of GIS/LIS that focus on the location of land features. This book seeks to bring special attention to those attributes that distinguish a modern ALRS from other location-oriented spatial information systems.

## 6.4   Attributes of a modern American land records system

A modern ALRS that serves all citizens is an object and a process that connects records and information about both the location and nature of land features and interests. Some attributes of existing systems require special attention to serve these demands.

Several criteria are used to identify the attributes that require special attention:

- They are not well developed in the existing American land records institutions.

- They suggest an especially lucrative, underused source of material.

- They promote citizen participation in land governance.

Creating an ALRS involves attention to the land governance process, including its underlying, long-held attitudes and practices. An examination of these processes is the basis for identifying the important ingredients of an ALRS and the attributes of these ingredients, which can be used to identify the substantive nature and extent of the records and information that constitute an ALRS.

A modern ALRS must include data and information with the following attributes:

- authoritative

- open

- complete

- connected

- timely

- institutional

Examples are presented to illustrate these distinctive attributes of a modern ALRS. These examples not only provide an understanding of the nature of these attributes but also indicate how geospatial technology can be employed to overcome the institutional barriers that inhibit connections between records and information about both the nature and extent of land features and interests.

## 6.4.1 AUTHORITATIVE RECORDS AND INFORMATION

*Authority*, as discussed in this book to describe land records and information, distinguishes and emphasizes the use of records and information rather than their production. Authority is established when the community, through its laws and processes, specifically gives status to a particular record or information as the basis for land planning and management. It is inappropriate, in this context, to assert that records and information are authoritative because of the nature of their source.

Within an ALRS, authority is what identifies and designates specific land records and information for priority inclusion in the system's database. This activity means selecting those data and information from among the vast array of available material that are most important to a specific community in the normal course of its land planning and management. This selection process imposes order on the chaos created by the technology-driven supply of undifferentiated, unexamined, and author-promoted data and information.

The designation of authority for priority inclusion in an ALRS does not give the selected material more legal status for land-use determinations than the selected material already has. However, this process does recognize the existing legal status that the material has already acquired in the land planning and management process. Attention to the already existing status and use of that status to give priority for inclusion is an important step in the allocation of scarce resources for design, implementation, and sustenance of a community land records system.

Many sets of land data and information contain a degree of this type of authority, such as legally designated maps used to determine which properties are eligible for low-cost flood hazard insurance. Generally, these datasets and maps have not been identified, designated, and assembled in an organized, public process and database. Reference to and use of these types of data and information gives direction and order to the building of a land records and information system. Modern technology makes these activities practical. The actions encourage efficient, effective, and equitable use of resources in construction and sustenance of a modern land records system. Finally, these actions are the basis for connections between the prioritized material in an ALRS and the ubiquitous activities of land planning and management.

A commonly understood, precise meaning of authority is crucial to the development of an ALRS. Authority often refers to an accepted source of expert information or advice. In a legal context, it often refers to the right to exercise powers, to implement and enforce laws, to exact obedience, to command, and to judge. In the context of government agencies, it often refers to the right or power of public officers to require obedience to their orders lawfully issued in the scope of their public duties. This elusive term can and does acquire a precise meaning based in the context of American land planning and management process.

Law and legal process dominate the practices of public land planning and management. The results of this process can be used to identify a range of authority for land records and information. The attributes of land records and information actually used to make land- and resource-use determinations become the priority material for an ALRS. This material, once captured, represents a significant, underused, and valuable source of land data and records that can be brought into a modernized system. Much of this material is scattered about the many agencies with statutory authority for administration of public land planning and management. Failure to capture and use this material shows the lack of development of American land records and information systems.

Agencies with statute-based land planning and management duties require records and information to support their plans, decisions, and actions. Some of the statutes are specific in their data and information demands. The materials used to satisfy these demands have a high degree of authority. Subdivision of land, approval of power plant construction, determination of landfill sites, a variance to a zoning regulation, and many other public and private proposals for a change in land or resource use require public approval. The statutes generate rules and regulations that describe the conditions necessary for approval of an application to change land use. These rules and the resulting practices often contain a demand for specific land records and information to be assembled to support the planning and management activity. These records and information are authoritative because they are required and because they are used in the planning and management process. They have acquired a well-defined, legal authority as the recognized basis for a land planning and management activity, and they can be identified and given priority in construction of an ALRS.

These data demands are familiar to those who participate in or observe land- and resource-use decisions by agencies. The federal National Environmental Policy Act (NEPA) demands a report by the deciding official, called an *environmental impact statement (EIS)*. This statement is required for every "major federal action significantly affecting the quality of the human environment" [National Environmental Policy Act (NEPA § 101) 1969, as amended]. The EIS contains the land records and information used by the agency official to decide which land records and information are left standing after the long notice and comment process.[5]

State and local governments also have statutes, rules, and practices that demand the submission of specific land records and information with a land-use permit application or similar government action. Applications for a subdivision permit or for low-cost flood hazard insurance are examples where the specific data requirements are designated by statute or rule. These specific land records and information demands are distinguished legally from data and information actually used to make determinations but not specifically designated in the agency's enabling legislation or administrative rule. An example of this latter type is the material used to determine whether a parcel or area is subject to a statute that prohibits activity within 100 feet of a wetland, but the statute does not designate the specific data or map to be used in the determination. In this latter case, the material actually used is often the result of a protracted ad hoc struggle in the administrative process.

Authority is an important aspect of all data. A precise definition and understanding of the nature of authority for all land records and information, such as those explored in the following sections, provides order to the chaos created by the vast array of land data and information generated and widely distributed by geospatial and information technology. The authority of data and information is what should be used to assign priority in an ALRS. The concept does not alter but recognizes the legal status of the data and information; however, the distinction could, in the future, become a basis for changes in the legal status of data and information.

*Authoritative title records*

The register of deeds receives documents that record the results of private negotiation regarding the transfer of land interests. Both property rights (rights, interests, and responsibilities) attached to parcels and the boundaries of these interests are described. Although the recording office provides no judgment of the document's validity, it does provide government recognition of the document's deposition in a public office. This recognition is sufficient to sustain the public-private partnership between the American system of land interest transfers and title assurance.

This system has several merits. It represents the establishment of a government office that imparts a degree of authority to important title documents. It is sufficient to sustain the dynamic public-private partnership that constitutes the American land titles institution regarding land interests: individuals and other entities pass title to real estate with a high degree of certainty about the allocation of land interests.

However, the system in regard to the location and extent of transferred land interests is not as certain. Documentation of the data and information are a major aspect of surveying and boundary law (Brown et al. 1986). The American system of individual surveys by a professional representative of one party associated with a boundary means that the description that appears in most transfer documents in the register of deeds has a low degree of legal authority. It has not been subject to the scrutiny associated with the adversarial process in American courts when real estate and surveying law are applied. An exception to this evidentiary and descriptive result is a court opinion following litigation between adversarial parties in a boundary dispute. The result should be given authoritative status and priority in an ALRS.

American survey law gives priority to on-the-ground physical evidence of the boundary of land interests, such as monuments and fences. The lines on maps or plats or in descriptions are indicators of the intent to place boundaries and markers at or near the depicted location. The locations of these boundaries are determined by the best evidence of actions taken on the ground to implement the intent expressed and recognized in the written material, such as intent to place monuments at or near indicated points (Brown et al. 1986). Evidence observed on the ground at a later time supports or fails to support the recognized intent. The plat, map, or description is an authoritative statement of the legally recognized intent to create parcel boundaries at or near the indicated places. The plat is not an authoritative statement of the parcel boundaries.

The American legal regime wherein land measurements are used to determine and assure boundaries between land interests has not kept pace with land measurement science. The priority of evidence for the location of boundaries remains essentially unchanged. This means that the best current evidence of the intentions of those who created the original boundary by measurement, demarcation, and description is current observation of the location of the original boundary monument or marker in its original location. The legal system, in the form of survey law, determines the priority for the current best evidence. This priority ranking does not change in concert with changes in measurement science. For example, witness evidence of the original location of the monument retains a high or highest priority, one not easily overcome by allegedly better measurements. Measurements must be between the correct points. Witness evidence is likely to continue to dominate arguments over which are the correct points to be measured. The position of geographic coordinates obtained from GPS measurements remains low because the issue is often whether or not the correct points for measurement have been identified (Brown et al. 1986).

The American system emphases of *private* professional opinions, descriptions, statements, and assurances of title and boundary rely on records and observations examined at the time land interests are transferred. These private opinions generally have no authoritative status and can be challenged before, during, and after they are made. This American system will likely prevail in the future despite efforts to introduce aspects of a title system that relies on government assurance of title and boundary. Despite this near-term probability, design of systems such as an ALRS should retain the option to move in this new direction.

Surveyors may assert that their measurements and measurement standards are very high and exceed those of others who measure the land. This work, summarized in title and boundary documents used to prepare parcel maps, may continue to be sufficient for identification and transfer of many land interests. However, it may not be sufficient to meet the demands for information in the course of twenty-first century land planning and management because it lacks authority.

## Authoritative assessment (fiscal) records

The local assessor (the auditor in some jurisdictions) does not impart authority to the records and information used to establish the value of parcels for tax purposes in a manner similar to authority imparted to deposited title records in the register of deeds. Most assessors are free to adopt practices that each regards as appropriate for assigning a parcel value for tax purposes. The parcel value acquires authority when it is designated as the legally recognized basis for a property tax. The parcel map does not acquire this type of authority.

The assessor's office became the home of the local government parcel map in most American jurisdictions. This was a creative, entrepreneurial action that evolved as assessors needed practical help in the efficient and effective completion of their mandated tax duties. The parcel map came to be crafted from data and information in title transfer documents deposited in the register of deeds. The description in these documents is often called a *legal description*.

The system of individual parcel boundary surveys significantly affects the attributes of local government parcel maps because individual surveys do not always fit together. The American emphasis on independent, successive surveys of the boundaries of a particular parcel, combined with independent surveys of adjacent parcels, often results in overlapping, multiple boundaries when all are considered. However, the general expectation is that a "clean" map should appear in the local government parcel representation. Powerful GIS technologies exist that assist with this capability. This technology enables a programmer or operator in the office that generates the local government map to use the technology to produce a clean map. Clearly, the local government parcel map generated in this way is a best representation of the boundaries without legal authority. A deeper concern with parcel boundary maps is that the boundaries only exist on the ground and that the documents and plans only indicate the location of boundary evidence.

Another way to look at a parcel map is that it could become an authoritative *index* to the authoritative document(s) that represents that specific parcel, rather than the authoritative, map-based representation of the boundaries. This status is achieved when the parcel map is used to establish authoritative parcel identification numbers (PINs), and these PINs are required to be used on many documents that describe both the nature and extent of land interests.

The absence of authority for the assessor's parcel data and information is best illustrated by the condition that the useful parcel map, assembled from various sources, has no authority in the title and boundary determination process. Some people may look to the assessor's parcel map for a general representation of parcel boundaries or for other useful purposes, such as maximizing travel time. However, those who are knowledgeable about the title and boundary determination process are well aware that the map has no authority in the determination of parcel boundaries.

## Authoritative natural resource data and maps

Consider three examples of natural resource land features with ambulatory boundaries. These examples are wetland maps, flood hazard maps, and nautical charts. These land- and water-related data and maps are given different degrees of authority in the land planning and management process. See figure 6.1.

1. **Wetland data and maps.** Wetland data and maps often are used when there is a desire to drain a wetland. This requires an application for a federal permit to dredge and fill in the nation's waters (Federal Water Pollution Control Act [FWPCA §404] 1972, as amended). The statute delegating authority to the agency or the agency's subsequent administrative rules do not specify a particular map prepared by a particular agency. This means that the quality of maps and data presented in the permit decision process can be contested. This is an ad hoc process in the sense that the data and information actually used to make the permit decision is determined during the legally required notice and comment portion of the administrative process. This contrasts with an a priori process wherein the map or data used to make the decision is legally designated and prepared prior to the administrative process. This legal designation gives authority to the map or data. Absent this authority, those who have the economic, political, or technical means to challenge the data and maps offered by others can acquire their own data and maps and assert ad hoc that this material be used for the decision.

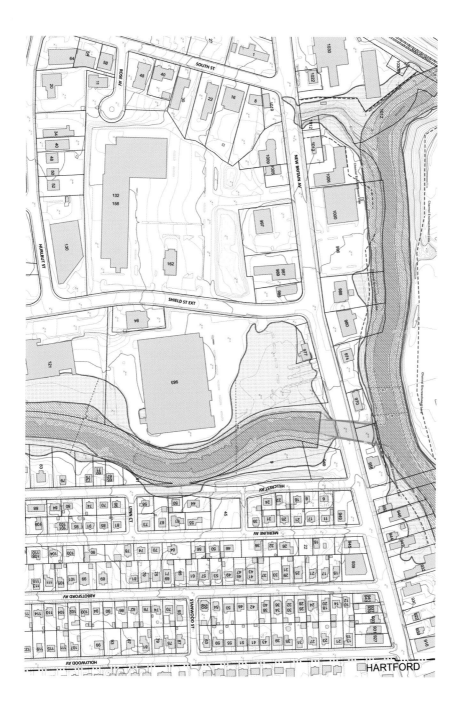

**Figure 6.1** FEMA-based floodplain determinations are overlain with property ownership parcel maps to graphically depict specific properties in West Hartford, Connecticut, that would benefit from flood insurance.

*Esri Map Book, Volume 26* (Redlands, CA: Esri Press, 2011), 58; courtesy of Applied Geographics Inc.

The ad hoc character of this process is captured by the lawyerly admonition, "Do not depend on government maps. With good science, the boundaries can be negotiated!"

Absent legally designated authority, wetlands data and maps initially submitted to the permit decider can be described as advisory, expert, or influential. If no one challenges the initially submitted map, often a government map, then it will be used to make the land-use decision. Wetlands continue to be drained in the United States when permits are granted to those who have the means to assert that their data and experts are better than the government's or other peoples' data and experts.

2. **Flood hazard data and maps.** Flood hazard data and maps are examples of material that does acquire authoritativeness for a specific land planning and management decision. This occurs in connection with the federal program that helps those with buildings in a flood hazard area to acquire low-cost flood insurance. The map used to identify the flood hazard area acquires authority for flood insurance purposes when a local jurisdiction designates a specific flood hazard map in a local land-use ordinance as the basis for determining the eligibility of existing buildings for federally subsidized low-cost flood insurance. This designated map is frequently the flood hazard map prepared a priori by the Federal Emergency Management Administration (FEMA). FEMA defines the map as "[T]he official map of a community on which FEMA has delineated both the special hazard areas and the risk premium zones applicable to the community" [Federal Emergency Management Act (FEMA) 1974]. This definition relies upon local legislative designation of the map as the authoritative basis for insurance decisions. Most, if not all, wetlands maps lack this authoritative designation.

   The statutory, a priori designation of a specific data and map means that the material must be used to make the land-use decision. The map acquires a high degree of authority. Claims about poor quality of the designated map may shift the venue for complaints about its quality from the ad hoc, notice and comment land-use drama dominated by lawyers to the a priori mapmaking venue more influenced by mapmakers, scientists, and engineers.

   Surveyors are sometimes asked to issue elevation certificates when the map may not be specific or accurate enough, providing a healing process for the authoritative data. These maps are used for a specific purpose—flood insurance—but have other good uses. However, these maps do not provide an exhaustive or complete dataset for all flood-related activities. These maps are authoritative for some very specific uses but not all uses. This distinction is important and should be a product of a modern land records system. It is also notable that these maps or data are often published as a web service, which means that anyone can connect to the data, introduce the data into their own systems, and use the data for their maps or spatial analysis.

The flood hazard map with high authority for low-cost flood insurance has a lower degree of authority when it is used for other purposes. The data and map have no legal status beyond their original purpose. The maps can be challenged ad hoc in the administrative process.

3. **Nautical charts.** Nautical charts issued by the US National Geodetic Survey, part of the National Oceanic and Atmospheric Administration (NOAA) under the US Department of Commerce, are specified by statute as the basis for navigation decisions by pilots licensed to guide vessels into ports.[6] For example, if the chart depicts an area within which a sunken ship lies, then the pilot is forbidden by statute from guiding a vessel over this area. If a pilot directs a vessel over this area and an accident occurs followed by litigation, then the pilot cannot assert in court, for example, that he and his fellow pilots have taken vessels across the area countless times and they all know through these experiences that there are no remains of the sunken ship. The only admissible evidence is the depiction on the chart. The chart has been given indisputable authority in the legal system as the basis for pilot navigation decisions. The chart also limits the pilot's liability if the chart is in error.

The chart may have other uses for which there is no legal authority. They can be used to paper the walls of restaurants and bars without concerns by the proprietor for misdirected sailing by drunken sailors.

*Authoritative infrastructure and other cultural records*

Many maps of the location of infrastructure or other cultural land features are not given recognized, legal authority for specific land planning and management activities. These include the collection of data and preparation of maps for non-owner-occupied parcels, organic farm fields, buildings with code violations, earthquake zones, curbs, water facility lines, and foreclosed properties. These data and maps are frequently used for many kinds of planning and management activities. They have status analogous to that of wetlands, as previously described. Their appropriateness can be challenged in a manner similar to that for the wetland data and maps. Many of these data and maps can be given authority for specific planning and management decisions analogous to the authority given to flood hazard data and maps.

*Authoritative administrative records*

Many jurisdictions use land-use control and similar legislation to regulate the nature and extent of public and private land- and resource-use activity. The statutes assign residences, industries, commercial enterprises, agriculture, and so on, to prescribed, sometimes exclusive zones. The legislation does not always prescribe the details of the data or map used to delineate the area subject to land-use controls. This lack of designated authority for data and maps occurs, for example, when a land-use-control statute limits land uses within 100 feet of a wetland. Nevertheless, the major legislative action of imposing a land-use control significantly alters the distribution of land interests in the community. That action needs to be better connected to the land records and information in a community so that all citizens are aware of the activity in a timely and appropriate manner so that they can fully participate in land governance.

There are many examples where land-use-control legislation does generate designated, authoritative data and maps. A map is given legal authority, for example, when it is generated specifically by agencies to implement and enforce exclusive zoning, depict building code violations, and so on. These designated, authoritative records and information can and should be distinguished from the spatial materials that have not acquired this important status. Modern geospatial technology makes this practical. All the designated, authoritative material needs to be assembled in a public land records and information registry similar to the register of deeds for privately arranged land interests.

*Authoritative land records: A basis for a modern ALRS*

American land records institutions make it difficult and sometimes impossible to assemble the data and information needed by individuals, groups, and officials involved in land planning and management. The existing land records institutions are particularly poor at connecting records and information about the location and extent of land features with material about the nature of rights, interests, and responsibilities associated with these features. At the same time, public land planning and management laws and legal processes require that both the nature and extent of land interests be considered when land- and resource-use plans, decisions, and actions are taken. It is not enough to know about the physical and cultural status of a place or area. It is also necessary to know who has the power to decide its use and how that power is exercised.

The laws and procedures of the public land planning processes impart a legally recognized authority to land records and information used to support plans, decisions, and actions. This ongoing activity can be used to identify records and information, if given the authority. Records and information given this authority in the normal course of land planning and management can be distinguished from records and information that are not. Authoritative records and information become the bases for land records and information files in an ALRS. Failure to clearly identify and distinguish those records and information that have authority for planning and management is part of the existing problem. The task of building an ALRS begins with recognition that the land planning and management process results in actions that give authority to records and information that affect the nature and extent of land interests.

Use of land records and information in land planning and management generates a range of authority. Table 6.1 describes this range of authority.

The authority of wetland data and maps used in the land planning and management process can be classified. This can be done for local, state, or federal agency administrative processes where data and maps are presented before an applicant can receive a permit to drain a wetland. This classification is indicated in table 6.2.

## TABLE 6.1   The range of authority imparted to land records and information

### High authority

Specific records and information designated by statute, administrative rule, or judicial opinion as determinative for a specific public or private plan, decision, or action. Flood hazard maps designated by local jurisdictions for use in the federal flood insurance program are an example. These records and information should be identified and become a part of an ALRS.

### Intermediate authority

Records or information commonly used by an agent of government (legislature, agency, court) as a matter of regular practice in the normal course of its duties, but without specific designation in a statute or rule. The use of an assessor's parcel map to indicate a parcel's area for property tax purposes is an example. These records could become a part of an ALRS if there is a common, recognized pattern of acceptance and use by those in the community who generate and rely on the material.

### Low or nonauthoritative authority

Records and information with an informal level of acceptance among community members who are interested in land-use decisions. A wetlands map used to indicate the extent of wetlands loss over a period of time during the course of debate about wetlands preservation plans is an example. These records and information can be held and used appropriately but should be distinguished from authoritative records and information.

**TABLE 6.2   A range of authoritative wetlands data and maps**

| Use | Representation | Authority level |
|---|---|---|
| Representation of the extent of wetlands loss in a state over two centuries | Small-scale map | Low |
| Preliminary indication of the general location of wetlands in a community | Small- to intermediate-scale map | Intermediate/ low |
| Evidence of the location of wetlands appropriate for application to specific parcels | • Intermediate- to large-scale map<br>• No designation by statute or rule as determinative | Intermediate/ high |
| Evidence of the location of wetlands on a parcel sufficient for the purpose of granting a permit to drain the wetland | • Large-scale map<br>• Designated by statute as determinative for the permit decision | High |

Wetlands maps with high authority could be designated for specific wetlands plans, decisions, and actions in a community for a specified period of time. This would be analogous to the designation by law of a flood hazard map as the basis for flood insurance decisions in a community. Whether this should be done for wetlands maps is a matter of policy in a community based primarily on the demands of wetlands planning and management and not on the supply of wetlands data and maps.

A range of authority can be developed for many or all land records and information used in the public processes wherein public and private rights, restrictions, and responsibilities are determined. Characterizing the degree of authority in this way allows designers, builders, and implementers of a modern, multipurpose ALRS to establish priorities for data collection and inclusion in their systems. This designation of authority to some data and not others can be used to establish priorities when investments are made in the development of data layers in agency thematic data layers.

Recent efforts to define authority fail to satisfy the demand for records and information that connect the location and rights aspects of land parcels. Authoritative data are often ambiguously described as data from a trusted source. Similarly, the Federal Geographic Data Committee (FGDC), organized and supported by the mapping division of the US

**TABLE 6.3   Federal Geographic Data Committee definitions of authoritative data**

| | |
|---|---|
| Authoritative data | Officially recognized data that can be certified and are provided by an authoritative source. |
| Authoritative data source | An information technology (IT) term used by system designers to identify a system process that assures the veracity of data sources. These IT processes should be followed by all geospatial data providers. The data may be original or it may come from one or more external sources, all of which are validated for quality and accuracy. |
| Authoritative source | An entity that is authorized by a legal authority to develop or manage data for a specific business purpose. The data this entity creates are authoritative. |
| Authority | In the context of public agencies, it is the legal responsibility provided by a legislative body to conduct business for the public good. |
| Authorization | The result of an act by a legislative or executive body that declares or identifies an agency or organization as an authoritative source. |
| Data steward | An organization within an authoritative source that is charged with the collection and maintenance of authoritative data. The term *data steward* is often confounded with the term *authoritative source*. |
| Trusted source and trusted data | A service provider or agency that publishes data from a number of authoritative sources. These publications are often compilations and subsets of the data from more than one authoritative source. It is trusted because there is an official process for compiling the data. |

Source: Federal Geographic Data Committee, 2008.

Geological Survey (USGS), has adopted a set of definitions that reflect a similar attitude regarding authoritative data. These definitions appear in table 6.3 (Federal Geographic Data Committee 2008).

The emphasis in these definitions is on the source of the data and information, not on its use. The implication is that a government agency, official, or professional expert is a trusted source. Therefore, the data or map should be trusted.

Each official, group, and individual professional can claim authority for the data and maps they produce. The material can be promoted as authoritative based on the expertise of those who produce the material. The problem with this designation of authority is that it relies too much on the conditions of data supply and not enough on the conditions of data demand. Land-use plans, decisions, and actions require data sanctioned by law or process for the specific purpose of public and private land-use actions.

Supply of land data must be matched to the demand for data in a land planning and management process. Land data authority is not based on the source of the data. It is defined by the data use in a particular context and purpose. The legally satisfactory use in a particular context gives data authority.

The distinction between authoritative land records and other land records is a basis for an ALRS that gives order to the land records chaos. The degree of authority acquired by land records and information depends on their role in the land planning and management process.

Land records and information can be designated by statute, administrative rule, judicial opinion, professional standard, or common practices as the basis for a land-use plan, decision, or action. The land planning and management process—governed by the statutes, rules, and practices established at all levels of government—provides the venue where land data and maps are presented, possibly attacked, diminished, or eliminated from consideration as the basis for land-use plans, decisions, or actions. Data and maps are evaluated and given status in that process, which can impart authority. The degree of authority depends on their role in the land planning and management process. The status of a particular map or data can be determined at or near the time of final action (ad hoc) or specified prior to the particular permit application (a priori). Data and maps only acquire authority through this process. This authority can and should be used to guide the choice of land data and record ingredients and establish their attributes in the development of a modern, multipurpose ALRS.

## 6.4.2 OPEN AMERICAN LAND RECORDS AND INFORMATION

Openness is a major attribute of land records and information in an ALRS designed to serve the demands of all participants in land planning and management. Openness determines who has access to land records and information used by government agencies in the normal course of authorized and mandated duties.

The topic of open data inspires contentious debate over questions such as the following: What records and information should be held in confidence? What should be done with material once it is released? How much should people and organizations pay for access and use?

The open-data controversy has a long history, beginning when records were largely in a paper form. To assure citizen access to records and information used by the federal government in the normal course its activities, Congress passed the Freedom of Information Act (FOIA) in 1966. The US Supreme Court interpreted the meaning of the statute in 1973 (EPA v. Mink 1973). The court found that the law's intent was to provide citizens with the right to obtain data and information used by governments in the normal course of their mandated activity from the hands of government officials sometimes unwilling to provide the material.

Each state has its own freedom of information laws, called *open records laws*. These laws vary, and each must be interpreted, and the interpretation must be applied to its agencies at the state and local levels of government.

The maturation of land measurement and computer sciences at the beginning of the 1980s led rapidly to the spatial information systems developed in the 1980s. Government officials at all levels and others sought to convince legislators and those who determined agency budgets that it was desirable for a variety of reasons to invest in these systems. Then and now, citizens and their elected officials ask how the investments can be funded. One suggested answer is that the investments can be successfully funded from revenues obtained from the sale of the data and information to those who wish to have and use the data. Agencies were encouraged and, in some cases, specifically told to authorize their data to be sold at prices determined by the agencies. The effort to establish conditions under which agencies sell government land records and information grew as records increasingly took a digital form.[7]

Jurisdictions were encouraged to sell their data at prices beyond those traditionally available under open records and freedom of information laws. Selling data at prices significantly beyond the reproduction costs called for open records and freedom of information laws to be changed or interpreted to exempt computer-stored spatial data and maps from the openness provisions that applied to other data. In a few states, statutes were changed to specifically exempt computer-stored land databases from traditional open records laws. All these promotional efforts asserted, or at least implied, that GIS/LIS development could not occur without access to significant data sales revenue.

Three decades of experience and observation with these efforts yielded the following:

- No large stream of system-sustaining revenues has emerged from sales of government land data.

- The pressure to have more efficient government operation has required legislators and officials to invest in all kinds of computer-based operations. These officials have not needed the incentive of data sales to build systems.

- Efforts to hold and sell the data used by governments in the normal course of their mandated and authorized activities impede the development of creative new businesses and uses based on these data.

- These efforts impede the ability of citizens to know what their governments are doing, which is fundamental to a participatory government.

Another aspect of openness is the confidentiality of government-held records and information. Here the issue involves general access to personal data such as income and health data and corporate data related to trade secrets. Concern about confidentiality has competing aspects. Consider, for example, health data. Many, if not most, citizens want personal health data held confidentially by public health agencies. On the other hand, citizens may sometimes suspect, for good anecdotal and other reasons, that the rate of an illness in an area is high and seek data to investigate their concerns. Lack of access to data impedes citizen participation in exposure and investigation of the issue. This is a significant dilemma in which a resolution is a matter that needs full public discussion. Confidentiality and openness are in a state of tension and require serious and complete discussion.

Withholding records and information from public access for data sales and protection of confidentiality defeats the basic purposes of the acts. General, convenient access to government records and information used by governments in the normal course of their duties is needed in order to know what governments are doing. This concern is basic to democracy and citizen participation in governance. The balance between openness and confidentiality must be carefully calibrated in a society where citizens insist on knowing what their government is doing.

Openness is essential in American governance. Americans expect to have a government whose actions and reasons for action are known. The historian Henry Steele Commager described the origins of this expectation. He noted that the "the generation that made the nation thought secrecy in government one of the instruments of Old World tyranny and committed itself to the principle that a democracy cannot function unless the people are permitted to know what their government is up to" (Commager 1972, quoted in EPA v. Mink 1973). Access to the records and information used by government to support their activities is one of the most effective ways to reveal what governments do.

Access to records means more than the ability to see the records. Citizens must be able to acquire the records in a form used by the government and have access to all the open material in order to understand what the records reveal about such issues as the fairness of assessments and the general distribution of police services.

The ability of citizens to acquire records in the form used by their governments is very important because of changing technology. The private sector can take this data, their own data, and data from other sources and create systems and applications and distribute the results in many new ways.

But these actions create a challenge. Paper maps can be time stamped and represent a situation at a particular time. Conditions are different when data and maps are provided as an electronic service. These services are live connections, and data are entered, revised, corrected, deleted, and otherwise changed. The history and details of this process, or metadata, are not now effectively communicated or distributed.

Government efforts to tightly hold their land records and information to sell it and manage its confidentiality restrict the ability of citizens to see and examine the data and information used by governments in the normal course of their mandated and authorized activities. This limits citizen knowledge of government activity and limits the extent of citizen participation in that government, which encourages government secrecy. Openness requires that sustaining data sales and confidentiality be balanced against the ability of citizens to know what their government is doing. Americans have traditionally favored the latter as represented by their long-standing lack of deference to officials.

Another argument for preserving data confidentiality is that citizens want their data to be confidential, so officials are merely acting on their behalf by withholding data. This eagerness of officials to act to protect citizens from the prying eyes of other citizens needs to be considered alongside the traditional desire of government officials to keep citizens from a full knowledge of what officials are doing. Traditional methods exist for protecting confidentiality other than a broad power placed in the hands of officials to withhold data. The federal Freedom of Information Act (FOIA) and state open records laws have confidentiality exceptions for those records and information regarded by the community as appropriate for withholding. This is illustrated by the set of exceptions to openness in the federal FOIA. The current list is summarized and is the same as those in the FOIA when it was first passed [Freedom of Information Act (FOIA) 1966]:

1. Those documents properly classified as secret in the interest of national defense or foreign policy.

2. Related solely to internal personnel rules and practices.

3. Specifically exempted by other statutes.

4. A trade secret or privileged or confidential commercial or financial information obtained from a person.

5. A privileged inter-agency or intra-agency memorandum or letter.

6. A personnel, medical, or similar file the release of which would constitute a clearly unwarranted invasion of personal privacy.

7. Compiled for law enforcement purposes, the release of which

   - could reasonably be expected to interfere with law enforcement proceedings;

   - would deprive a person of a right to a fair trial or an impartial adjudication;

   - could reasonably be expected to constitute an unwarranted invasion of personal privacy;

   - could reasonably be expected to disclose the identity of a confidential source;

   - would disclose techniques, procedures, or guidelines for investigations or prosecutions; or

   - could reasonably be expected to endanger an individual's life or physical safety.

8. Contained in or related to examination, operating, or condition reports about financial institutions that the SEC regulates or supervises.

9. Documents containing exempt information about gas or oil wells.

Those unfamiliar with the legal process look at this list and conclude that an official can easily find a reason or craft an interpretation that allows withholding data; however, the US Supreme Court and most state supreme courts take the approach that openness is to be interpreted broadly, while exceptions are to be interpreted narrowly. This provides a strong bias in favor of openness when a dispute reaches the courts. Finding and keeping a balance between openness and exceptions to openness is a constant and dynamic process. A starting point for a discussion of a balance between confidentiality and openness is the observation that secrecy is for people, while transparency is for governments.

Unfortunately, this statutory and judicial bias in favor of openness does not prevent some officials from resisting claims for data. Sometimes, this is done because officials are uncertain of the status of an exception. Other times, it is done knowingly and deliberately under the assumption that the requester is unlikely to undertake the arduous task of asserting her right to the data in court. The legal maxim here is, "there is a right, and there is remedy." Sometimes, officials put a citizen to the arduous task of a court remedy even when the official knows that the citizen has a right to have the data.

Another aspect of openness, not revealed by the discussions of data sales and confidentiality, is the active effort by some governments to provide data and information to their citizens. Modern geospatial and information technology are used to do this when, for example, an agency posts records and information on its web pages. This is an action to be encouraged wherever possible.

This activity constitutes a one-way portal into government records and information. It gives the agency official the power to decide what records and information, and in what format and other conditions, will be provided to citizens. If a citizen wants more data than that provided on the web page or wants the entire database used to generate the material that appears in the government web page, a negative response from the official may follow: "There is all this data on the web page. What more do you want!" The citizen may want to investigate the fairness of the entire parcel valuation process in a community and need all the data to do so.

Open records and freedom of information acts consider the extent of a citizen's ability to acquire the records and information used by government agencies in the normal course of their duties and responsibilities. The emphasis is on what of the government-held material is available to citizens, under what conditions, and how the satisfaction of a request for data and information is executed. Modern technology expands the scope of these considerations.

The attribute of openness also includes the ability of citizens to generate records and information and introduce these materials into agency processes and databases. The ubiquitous notice and comment process is one example of this.

Another mechanism emphasizes active efforts by officials to not only receive the material gathered by citizens but also to establish recognized procedures for incorporation of the material into the agency databases and for giving the material authority. To do this, the agency must establish a process to receive the material, examine the quality of the material for use in their mandated duties, and formally accept the citizen-proffered material by granting the material the authority necessary for its use in agency land planning and management. The material becomes a part of the records and information designated for use in making plans and decisions and taking actions.

Examples of this kind of activity exist. For many years, the Coast and Geodetic Survey (now the US National Geodetic Survey), part of NOAA, has had a process that responds immediately to data and information brought by citizens to the agency about the status of objects or conditions displayed on nautical charts. Citizen reports of discrepancies are evaluated by the survey. If they are found accurate, then the changes are reported in the authoritative *Notice to Mariners*. This notice has legal authority in that pilots are required by law to execute their work with knowledge of the changes in the nautical chart reflected in the notice.

Another widely represented example exists of an established, active procedure wherein an agency responds to input of data from an external source and changes a database and map. This occurs in the administration of flood hazard data submitted by a local government to FEMA in the US Department of Homeland Security. A local government can submit this data, allegedly better than the material held and used by FEMA to make flood hazard maps, with the expectation that the material will be considered as a replacement for the authoritative flood hazard data and maps used by FEMA for decisions about low-cost flood insurance.

This active response process wherein citizen input to a receptive data-authority-creating agency has several benefits. One benefit is that it demonstrates active agency response to citizen and other external records input. Another is that it provides an additional, new source of land records and information at a time when resources for data collection and maintenance by agencies are limited. Finally, the active agency process creates a two-way, open information portal with the potential to increase positive citizen attitudes and practices toward government. Active government response by agencies to citizen input creates a means to increase citizen participation in land governance. See figure 6.2.

### 6.4.3 COMPLETE AMERICAN LAND RECORDS AND INFORMATION

Modernizing American land records and information institutions requires attention to the attribute of completeness in all its aspects. One aspect is the need for complete records and maps of the surface boundaries of all parcels in a community, the location of gas lines, and other land features.

An ALRS should also include complete land records and information needed by the community in the normal course of its land planning and management. Many land features are included. The choice can range from wetlands to gas lines and from earthquake-prone areas to telephone pole locations. The choice depends significantly on the physical and cultural conditions of the community, such as its urban, suburban, or rural character and its physical status of the land and its resources. This aspect of completeness again draws attention to the local character of land and land governance in America.

When considering completeness of content in an ALRS, one should consider the following question: What land features, and what attributes of these features, should be included in the ALRS?

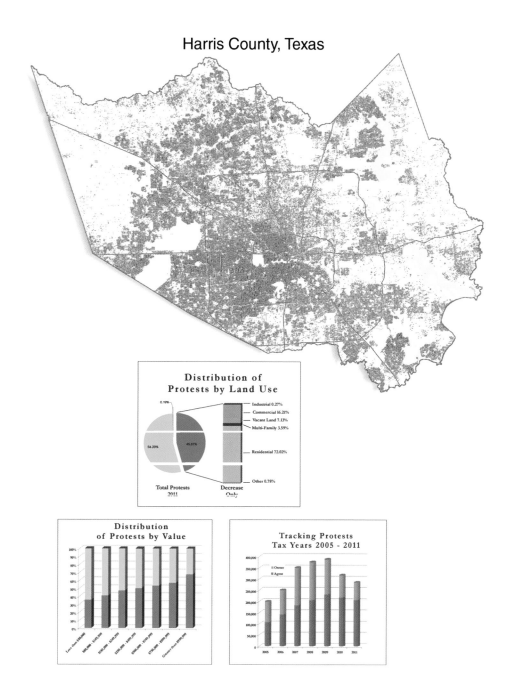

**Figure 6.2** The Harris County Appraisal District in Texas has a protest process that resolves property owners' disagreements over property values. The charts show protests by type of land use, by value, and by years.

*Esri Map Book, Volume 27* (Redlands, CA: Esri Press, 2012), 52–53; courtesy of Harris County Appraisal District.

A difficult problem for modernization of existing land records and information systems is that of identifying those important land features in the local community for which the set of rights, interests, and responsibilities need to be connected to the location of the features. Land records and information in a community are not complete if, for example, it is not practical for citizens to determine the distribution of mineral rights in a community subject to the potential for hydrological fracturing (hydrofracking), while wealthy or powerful companies alone have the means to make this determination.

What publicly and privately created rights, restrictions, and responsibilities shall be a part of an ALRS such that the system can be called complete? The answer depends on the set of rights, restrictions, and responsibilities that the community regards as important for management of land and resources in the community. It depends on the number and type of these management decisions that the community currently makes. It also depends on the set it expects to make in the near and distant future. The ability to identify the number and type of these decisions and to measure the efficiency, effectiveness, and equity of an effort to assemble the anticipated records and information in an ALRS is a challenge for any community. However, it is appropriate, if not necessary, that the effort be a part of a community's twenty-first-century planning and management. Plans for community land development can and should include plans for the records and information that facilitate the proposed development.

Another aspect of completeness involves records and information collected in the past, which are important because they often contain data and information about past conditions that are relevant in current land planning and management activity. For example, PLSS records contain not only measurements and marking of parcel boundaries but also observations of land and resource conditions at the time of the initial survey work. These observations can be important in current public and private considerations. These considerations can involve actions designed to preserve existing plants and trees that are alleged to be remnants of old-growth flora. Many historical records in addition to PLSS records exist, but they are difficult to locate even if their existence is known. However, the efficiencies of modern geospatial technology suggest that a community can and should identify appropriate historical records for digitization and inclusion in the array of land records in a modernized ALRS.

Complete ALRS data can include a document that indicates a potential public land interest action, such as a land-use control. For example, proposed legislation to control the use of land within a specified radius about a proposed wind turbine location can be connected to an ALRS. Application for a permit to construct a wind turbine at a particular location can be followed by use of a parcel map to identify all areas affected by the land-use-control legislation and notification distributed to all affected and interested landowners in the site vicinity before final action on the permit application. Notification of the introduction of a land-use-control bill or of a permit application combined with reference to the bill or application allows interested parties to quickly locate and read the proposed legislation or existing legislation that governs the permit application. Modern technology makes it efficient to complete the public record in this way.

Consideration of historical, existing, and potential land records in a complete ALRS makes it possible for people and organizations with a keen interest in their circumstances to follow a daily inquiry about the status of their stock portfolio with an inquiry about the status of their land and land interests with the expectation that a reasonably complete answer is easily obtained.

### 6.4.4 CONNECTED AMERICAN LAND RECORDS AND INFORMATION

The technologies previously discussed also promote the ability to generate useful data and information about parcels, such as PINs, which uniquely identify land parcels. It is easy and convenient to generate a PIN that contains valuable parcel information, such as the unique latitude and longitude or the geographical coordinates of a point in the parcel. The unique PIN is the means to make connections between parcel maps and the many parcel records that contain information about the nature of land interests associated with the parcel such as title transfer documents, a public action granting a variance to a zoning designation, or a notice of a building code violation.

The ability to generate PINs with valuable data and information and to distribute these PINs easily has not resulted in the widespread use of PINs to connect parcel location data and maps with documents about the nature of parcel land interests. It all seems so easy to use PINs to connect records of the nature and extent of land interests. Despite the development of geospatial and information technology, the 2007 NRC report, *National Land Parcel Data,* indicated that little progress has been made in the widespread use of a common PIN on a variety of land records in a community since the NRC report, *Need for a Multipurpose Cadastre,* described this role for the PIN as the basis for making these connections between land records.

This aspect of agency independence is an example of thinking within an existing institution. It results in the circumstance previously described wherein the ability to generate and distribute an important land records attribute, the PIN, is not followed by the widespread assignment and use of the number by a variety of land records agencies. The failure to do this makes it hard if not impossible to assemble a set of documents that reveal the status of land features and the land interests that attach to these features at the time and place needed for their use in land planning and management. The costs of overcoming these difficulties are imposed, not on the agencies that handle land data and records, but on those in the community who need to assemble the information package that is specifically required in the land planning and management process. These costs to the community can be demonstrated, even if their full scope can only be partially revealed. These costs are described in chapter 7.

Coordination of standards among agencies for both data collection at a particular area, site, or parcel and the use of PINs for both parcel maps and parcel-related documents can enhance the connectedness of authoritative data and generate benefits to the community of land data and records users.

## 6.4.5 TIMELY AMERICAN LAND RECORDS AND INFORMATION

Timeliness of land data is an important attribute of an ALRS. *Timeliness* usually triggers concerns about the currency of data about the location and condition of land features, such as roads, wetlands, flood hazards, potential urban gardens, and telephone poles.

Timeliness also has another, related aspect involving arrangements between public and private parties that result in such property rights as scenic, utility, and similar easements. The landowner relinquishes the right to change the scenic nature of the land, while a public entity (e.g., a public park agency) or private (e.g., land trust) group acquires the duty to sustain the easement conditions for a long time, even in perpetuity. The landowner retains control of most of the other land-use rights. When these other rights are sold, the easement holder needs to be notified of this change in a timely manner so that there is assurance that conditions are maintained. At a minimum, the easement holder wants to make sure that the new landowner is notified of the easement conditions before untimely changes in land use by the owner defeat the easement conditions.

The lack of timely notice to others of the common action of property transfer has costly consequences. The register of deeds receives a copy of the land transfer document. The office records not only the document but also makes a note in its grantor/grantee index, tract index, or both. The office has created this important, legally significant set of records. It is not common for the register of deeds to notify anyone about the records it has created, or in this case, that a transfer has occurred. Failure to do this causes others to maintain expensive actions. For example, the Saint Croix National Scenic Riverway of the US National Park Service holds about 1,300 scenic easements along the river in many counties and in two states. Parcel ownership changes regularly. The riverway administrator is eager to inform the new owner in a timely way of the easement restrictions. In order to do so, an agency official must devote significant time to the task of keeping the riverway's record of owners up to date. And, this may not be enough time (Saint Croix National Scenic Riverway administrator, personal communication, September 2011).

It is not common for a register of deeds to provide this type of information to others, even though the register of deeds commonly creates a notice of a transfer record within its indexes of documents. It is very easy for the register of deeds to transfer this information using modern information technology once the register's operation is digitized. This activity is not a part of thinking within the traditional register of deeds institution. It is a part of an ALRS.

Other land information institutions—such as the assessor's office, local planning or zoning office, or engineer's office—can participate in this modern distribution of timely information.

An important new technology affecting both completeness and timeliness of land records in a modernized ALRS is the unmanned aerial vehicle (UAV). This instrument can collect data about the status of land features accurately, rapidly, and inexpensively over a small area. An individual or local agency can obtain land data and information instead of data from an airplane or satellite. This makes it possible for government agencies to add a powerful new tool to their land data-gathering arsenal of equipment, reducing the cost and time necessary to obtain the data. Governments and research institutions are uniquely able to take advantage of this new data collection technology in America. Current law that prohibits use of these devices by private individuals and organizations does not apply to government agencies and research institutions. However, privacy concerns exist among some members of the public. Development of this technology involves a balance between public and private interests and concerns.

## 6.4.6 INSTITUTIONAL AMERICAN LAND RECORDS AND INFORMATION

One of the major deficiencies of the existing ALRS is the absence of a central repository for the documentation of public actions that alter land interests in a community. This repository could take the form of a registry for public action documents analogous to the register of deeds for private-action documents affecting the nature and extent of land interests (e.g., deeds, mortgages, and liens). Alternatively, modern information technology makes it possible to create a virtual repository for assembly of documentation of these public actions based on computer-based connections between identified, appropriate records in various public agencies.

The problem is primarily institutional. It is not common to find a central registry of important public land interest action documents in a community. The challenge is to gain support for and implement a set of appropriate standards and procedures for the registry. This must be followed by actions, supported by administrative procedures, that define the important public actions that alter land interests (e.g., zoning controls and permits), identify the appropriate documentation, and provide for the transfer to or coordination of these documents and their contents for the registry.

Modern information technology makes this activity possible, practical, and efficient. The use of parcel identifiers encourages connections between parcel maps and parcel-specific documents.

The provision of a registry containing a full range of connected, useful land records encourages the expansion of the nature and scope of land data and records and their analysis for purposes of land planning and management. American attitudes and practices suggest a major role in the analysis, opinion, and assurance of the meaning of these public actions by private companies in a symbiotic relationship similar to that associated with the privately created documents recorded by the register of deeds and examined by abstractors, surveyors, title attorneys, and title insurance companies.

Twenty-first century land planning and management and geospatial technology development suggest strongly that it is time for this public registry as a part of a modern American land records institution.

## 6.5    Conclusion

This chapter identifies the ingredients and attributes of a modern, multipurpose ALRS.

Modernization of existing American land records and information systems and institutions emphasizes the fundamental importance of the needs for land data and records in the land planning and management. The existing American land planning and management process, and the fundamental attitudes and actions that it represents, provide both ideas and forces for modernization. The primary goal of modernization is an enhanced ability of a land records system to satisfy the needs of all parties who participate in determining the use of land and its resources.

The list of important ingredients in an ALRS identified in this chapter is based on those that are important in existing systems and for which modernization of the existing arrangements are needed. The ingredients include private title records, assessment, records of public actions affecting land, and data about land features.

This chapter has also identified the important attributes of these ingredients that require particular attention in modernization efforts, including the following:

1.  Authoritative data and records in the administration and governance of land and its resources.

2. A connectedness between independent land records institutions and databases.

3. Open data and records in the sense of a two-way information portal wherein needed material can be both obtained and contributed by individuals and groups in a well-developed, active, and responsive administrative process.

4. Complete records of both the nature and extent of land interests and features such that a complete record of public and private interests can be ascertained.

5. Timeliness in the sense that historical, current, and proposed data and records are available for current and future land planning and management.

The need to connect both the nature (public and private land rights, restrictions, and responsibilities) and extent (location) of land features interests has been a dominant theme throughout the effort. The land parcel has been the dominant basis for establishing that connection. The local level of activity has also been a focus of attention in accord with basic, long-established American concepts of land, land records, and governance.

# Notes

1. *American Heritage College Dictionary*, 4th ed., s.v. "institution."

2. A distinction is made between collective actions in the form of statutes that regulate activities, such as those that control the release of pollutants from a coal-fired power facility, and those statutes that delegate authority to decide whether or not a facility can be constructed at a particular site. Most federal environmental statutes that regulate pollution emissions from a facility leave the choice of a new facility's location to state or local officials.

3. Use of the term *cadastre* to refer to a land records system that encompasses a broad array of material about both the nature and extent of land features and interests is consistently resisted in America. If the term is recognized or used, it is in reference to a parcel map, called the *parcel fabric,* or the parcel boundary data and information that support the parcel map.

4. About a dozen states have passed enabling legislation that makes it possible for the state or a local jurisdiction to establish an agency that operates a cadastre or Torrens system that provides a government guarantee of title and/ or boundary. However, the number of places with significant activity is limited. Massachusetts has a state-level land court. Cook County (Chicago), Illinois, and Hennepin County (Minneapolis), Minnesota, are the most active. The national aspects of the American real estate market and of title examination and insurance mean that buyers and sellers continue to rely on private title assurance for most real estate transactions in all jurisdictions, even in jurisdictions that

have Torrens title systems. A summary and review of the arcane subject of Torrens title in America, written more than thirty years ago, is still valid in its fundamentals (Shick and Plotkin 1978).

5.  The NEPA requires federal agencies to prepare a report to accompany "major federal actions significantly affecting the quality of the human environment." Five specific topics must be discussed. This report, the EIS, is a major feature of agency activity. See 42 USC § 4332(C), NEPA § 101. The statute delegates to the agency the power to establish detailed rules and practices that determine what data and information is expected in a normal situation. For example, the US Department of Transportation has rules and practices, known to all interested parties, that specify what data and information is needed in the EIS for a standard highway or bridge project. Some states have similar laws for their state agencies. See, for example, the Wisconsin Environmental Policy Act (WEPA) (Wis. Stat. §1.12 and Wis. Adm. Code NR 150). Other states, without an overriding statute applicable to all state agencies and actions, have specific data and information demands in the specific applications for power plants, wind farms, subdivisions, state highways, etc. These demands are analogous to those in the overriding federal and state acts. All these statutes, rules, and practices give authority to the land records and information that the agency finally selects as the informational basis for its final action.

6.  Traditionally, changes in conditions represented on nautical charts were updated by means of a published, weekly notice, the *Notice to Mariners* (NM). The contents of this notice had the same legal status as the information in the charts themselves. The Coast and Geodetic Survey (now the US National Geodetic Survey) exhibited its willingness to receive, evaluate, and use any observations brought to them by citizens by the act of establishing a formal process for these actions. Citizens often made it a part of their boating pleasure to identify, locate, and report these changes to the Coast and Geodetic Survey. The traditional legal status of the NM is retained in the modern digital age. See the Nautical Charts and Publications web page, http://www.nauticalcharts.noaa.gov/staff/chartspub.html.

7.  In the early 1990s, it was common for inexperienced agencies to hire private consulting companies to help with development of agency land and geographic information systems. At this time in the early development of these systems, agency officials faced the daunting task of convincing elected officials and agency managers to invest in the new, largely unknown systems. One assertion presented to those with the power to allocate development funds was that investment costs could be substantially recouped by the agencies in the form of revenues from the sale of the new, closely held, computer-stored data and records. No significant stream of supporting sales revenues has emerged in America. Not only does an effort to hold data closely inhibit system development and use, but it also inhibits the ability of citizens to know what their governments are doing. The struggle of agencies to hold and sell the data and records they use to execute their mandated and authorized activities continues in some places despite the absence of evidence to support a stream of revenues. The continuing sustenance of open records by state and federal courts has not suppressed efforts to hold public records tightly and for a price. Organizations now support the resistance to the principle that citizens in a democracy have the unfettered right to see and fully examine agency records that document what a government is doing. One example is the Open Data Consortium project.

# Chapter 7
## Incentives, barriers, and prospects

Incentives and barriers to development of a modernized, multipurpose American land records system (ALRS) are economic, institutional, technical, and social. Incentives include performing tasks more efficiently and inexpensively, generating new and better products from the same underlying data and information, and providing greater access to data and information. Barriers include overcoming traditional American attitudes and practices involving land, land records, and land governance; institutional resistance to change; funding; and lack of full deployment of geospatial and information technology.

Barriers to change are associated with long-established preferred attitudes and practices. Public and private land records and information institutions are entrenched in a world of traditional activities. Proposals for change confront statements such as, "That is how it has always been done," or "That is a local political practice not easily changed," or "That is my vested interest." Science and technology can drive changes in attitudes and practices, even in long-vested interests; however, the nature, extent, and pace of change, even technology-driven change, is dependent on social, political, and economic forces.[1]

## 7.1    The economics of information

Economics is a force for change that often overrides other forces. Economic forces can be both an incentive and a barrier for change in traditional land records and information institutions.

Investments in land records and information systems require special considerations because information is a peculiar economic commodity (Mackaay 1982). Attributes that make it a difficult object of study include the following:

- Examination of the material is tantamount to use.

- Use does not deplete the amount and physical condition of the material.

- If information is released, it is hard to control its spread and use.

Economics is about the allocation of scarce resources. Land and its resources, as well as labor, capital, and information, are part of this allocation. Allocation implies choice. Choice implies a condition of uncertainty, a disability to be overcome. Information is an antidote to uncertainty. Information is receivable only when there is uncertainty. Uncertainty can be reduced to a point that people are willing to absorb the remaining uncertainty and to act.

Costless searches imply no limit to the information search. If cost is attached to information gathering, an exhaustive search is not optimal. Data and information are worth acquiring only to the point that the decision maker is sufficiently satisfied with the level of uncertainty such that informed action or inaction is the result.

Generating information is not the same as using it. The use of information to reduce uncertainty in management decisions implies a limit to the amount of information search. The prevailing aphorism is, "Not a search for the best information but the best information worth searching for" (Mackaay 1982, 10). Therefore, the level of investments in information search is a social as well as technical matter.

How does the decision maker know at what point uncertainty is reduced sufficiently that action can be taken? It depends on the circumstances. If the major plan, decision, or action is about how to use land and its resources, then there is one level of required uncertainty in the records and information. If the major concern is what accuracy shall data have without appropriate regard to its use, then the answer may be different.

## 7.2   The economics of land records systems

Preferred community attitudes and practices in regard to land planning and management establish the context in which to measure the nature and extent of investments in land records and information system. Although it is difficult to identify and measure the intangible values people assign to land and its resources, the land planning and management process provides observations that can be used to overcome the difficulty. Laws and legal processes governing the role of land records and information in land-use decisions reflect long-established attitudes and actions about how rights to land and its resources are made and what data are to be used, irrespective of when they are made.

The use and value of investments in land records and information systems and their products are usually described and measured, and most easily identified, from the perspective of the organization that makes the investments. The private or public organization asks the basic economic question, what are the identifiable and measurable benefits of investments in a new or altered system compared to investments in traditional actions and products?

Answers to this question can be confined to the observed activities of the investing organization or, even more narrowly, to the activities of the subagency directly involved with the land data and information. Answering this question in this narrow way is thinking within an institution. Investments in land records and information by a public agency whose primary duties involve land planning and management require thinking more broadly than one restricted to that of the investing agency. While taxpayers expect an agency to execute its delegated tasks efficiently and effectively, they also expect agencies to provide public benefits in ways not confined to the agency alone.

It appears that many investments in land records and geographic information systems over the last several decades were made without a priori identification of measureable benefits. Instead, system promoters asserted that investments could be recouped by subsequent sales of data to eager buyers. At the time, no measurable data supported this assertion. Nevertheless, the bold assertions were assembled and given wide circulation.[2]

The effort by agencies to hold spatial data and information for sale at prices beyond reproduction costs continues. This effort, inconsistent with open records laws in most states, often lasts until some organization or citizen sues the agency. This litigation consumes valuable public resources. In a series of cases in California, counties have been challenged when they refuse to convey what are often called *GIS databases* or *landbases*. The California Supreme Court has consistently supported the traditional open records principles. In the most recent case, Orange County was told that the court believes that "the public records exemption for 'computer software,' a term that includes computer mapping systems, does not cover GIS-formatted databases like the Orange County Land base at issue here." Orange County must respond to Sierra Club's request "in any electronic format in which it holds the information at a cost not to exceed the direct cost of duplication" (Sierra Club v. Orange County 2013).

Although measurable benefits may not be easily found within the government agency that makes the investments, society's expectations of accessibility to and use of records and information evolve, and evolve quickly with technology, altering governments' tactical functions and the community.

Many legislators, officials, and others quickly realized that the systems were necessary in order to execute their duties even without a tangible measurement of benefits.

It became clear that land planning and management agencies would not be able to execute their mandated or authorized duties without the land records and information systems. The need to execute mandated duties was significant, not easily subject to tangible measurement, and generally determinative in convincing many otherwise reluctant legislators and officials to make the investments. As the mechanics of executing government functions change, citizen expectation of access to data and information with these new systems change as well.

The challenge for those who promote modernization of land records and information systems is to expand the scope of thinking about the domain of activity wherein the benefits of investments are found beyond that of the investing agency. In regard to this type of approach, Albert Einstein is reputed to have said, "Problems cannot be solved by the level of awareness that created them."

Attention to this broader domain can also expand the extent of measurable benefits as well as clarify the nature of what is not measurable in the use and value of land records and information in land planning and management. Those who use land data for planning and management often assert that investors will benefit from the ability to do things that could not be done before. However, CEOs, elected officials, or others who are notoriously parsimonious with expenditures, often require a more precise answer to questions about the benefits of investments in a modernized ALRS.

# 7.3    Benefits of investments in land records for planning and management

Development of a model of the benefits of investment in land records and information for community use in land planning and management requires several activities. First, distinctions must be made between underlying data, information, hardware, and software components in the agency and the nature, extent, and form of land records and information products actually sought and used by participants in the land planning and management process.

Second, the unusual economic aspects of information must be considered. For example, we must consider that examination of information is tantamount to use and that the use of accessible public records and information is hard to control.

Finally, the contributions of land records and information to the land planning and management process are hard to separate from other influential factors. These other factors include the local attitudes and practices of individuals and groups at or near the affected land parcels (Moyer 1975, 1980).

The constant challenge to economists is to expand the domain of what is measurable and to clarify the domain of what is not measureable (Duchesneau 1982).

### 7.3.1 CLASSIFYING BENEFITS

The benefits of an ALRS, both measureable and not measureable, can be classified. They are defined and placed into one of three categories: efficiency, effectiveness, and equity.

*Efficiency* results when system technology makes it possible to execute traditional data management and mapping activities or products at a reduced cost, or to generate more products with the same resources, or complete tasks more rapidly.

*Effectiveness* results when the system makes it possible to generate new products from the same underlying data and information. These can include products that could not be generated without the new technology or products that could have been generated using the old technology but weren't because the cost or time involved were regarded as prohibitive. New products can be designed to meet the needs of those for whom traditional products are unavailable or inappropriate.

*Equity* is the benefit that results when more citizens and groups participate in land planning and management. These larger, community-oriented benefits have been labeled as equity (Kishor et al. 1990; Cowen 1994), decision making (Pinto and Onsrud 1995), societal benefits (Clapp et al. 1989), democratization (Lang 1995), and equity/empowerment (Tulloch and Epstein 2002).

One aspect of the equity benefit is that the number of those who use or seek system products in the community is much greater than the number of users in the organizations that generate the data and records. The set of users includes lenders, attorneys, developers, real estate brokers, land information companies that provide information services to landowners and buyers, agency officials, and power companies. It also includes environmentalists, community groups, and individuals who want to know what they or others can do with land and its resources. The number of individuals and groups who can use or have access to geospatial technology and data is increasing. They can analyze the data, form their own opinions and representations, present their opinions and representation to decision makers, and participate in the planning and management activity.

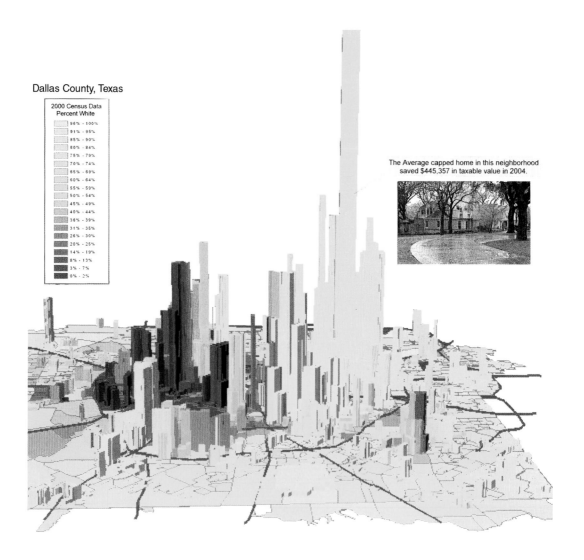

**Figure 7.1** The Dallas Central Appraisal District in Texas used a series of two- and three-dimensional maps to demonstrate that the largest amount of property real estate tax caps and resultant tax savings were accruing to the unintended—the wealthiest residential property owners.

*Esri Map Book, Volume 21* (Redlands, CA: Esri Press, 2006), 58; courtesy of Dallas Central Appraisal District.

Equity benefits include those that result when the poor and powerless are better able to use the existing system to more fully represent their views in a nuanced and sophisticated way alongside the wealthy and powerful who have advantages when records and information are scattered and disorganized (figure 7.1).

The label of equity for these benefits is appropriate because modern land records and information systems have the realizable potential to increase the nature and extent of land-related decisions that reflect a representative segment of the community. Achievement of this benefit is characterized by increased citizen participation, engagement, and empowerment. It represents democratization of system use and governance.

There is a timely, special opportunity for developers of a multipurpose, modernized ALRS. The current, popular political culture exhibits considerable hostility to government generally and to national government particularly. A modern ALRS that operates as a two-way information portal serves all citizens and groups and provides the means to enhance citizen participation in land governance.

## 7.3.2 OBSERVING BENEFITS

The first task is to identify the venues where products are used. The second task is to develop a model of how those products are used within those venues.

Efficiency benefits accrue primarily to the subgroup within an agency or organization that generates spatial data and information products from land data and information. An example is the mapping and GIS/LIS section of a planning agency or assessor's office.

Effectiveness benefits accrue to the agency or organization whose work depends on the subgroup that generates the products. An example is an assessor who depends on a parcel map and parcel data for effective and defendable parcel valuations (valuations that can be successfully defended when challenged by owners).

Equity benefits are found among the broad community of citizens, agencies, organizations, developers, companies, and others who have an interest in land planning and management. They are drawn to the land records institutions that provide data and information. Some support the existing land data chaos because it would undermine their assertion that they have the best data and can provide the best data services. Others support institutions that provide authoritative data to all in an administrative process wherein land-use decisions are made with input from all interested citizens. Equity benefits accrue to a community that recognizes a need for increased citizen participation by all interested parties in land plans, decisions, and actions (figure 7.2).

## 7.3.3 MEASURING BENEFITS

Efficiency benefits are the easiest of the benefits to measure quantitatively, using measures associated with benefit/cost determinations. Cost avoidance dominates the analysis. Efficiency benefits are measured by comparing the cost to collect, assemble, analyze, and distribute data and their representations using new and old methods.

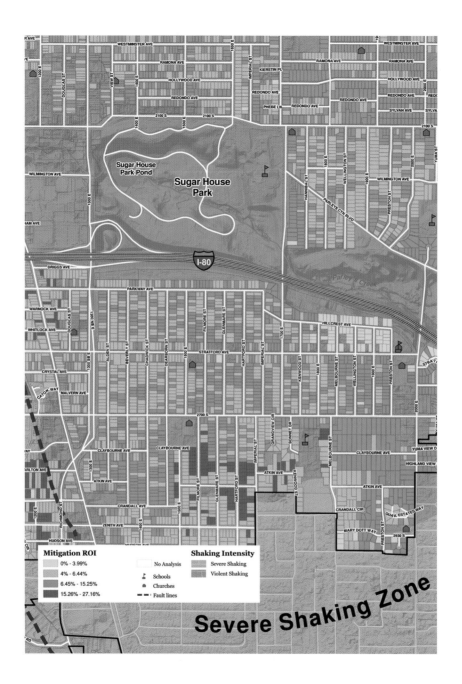

**Figure 7.2** Estimates for casualties, building damage, and economic losses were conducted for earthquake mitigation studies in Salt Lake City, Utah. When combined with parcel geometry, all properties were analyzed and identified to determine the benefits of mitigation measures—known as the *return on investment (ROI)*.

*Esri Map Book, Volume 25* (Redlands, CA: Esri Press, 2010), 61; courtesy of FEMA Mitigation Directorate.

For example, an assessor can compare the costs of parcel mapmaking before and after introducing new geospatial technology.

Effectiveness benefits are determined by comparing the cost of executing an agency's or organization's mandated or authorized duties with and without the new products of the geospatial technology. Attention extends to the full array of agency or organization activities supported by the land information system.

The key to measuring all effectiveness benefits is a complete understanding of the agency's or organization's basic duties and activities. An example is an assessor who uses a new mapping system to better direct the movements of employees as they do house inspections in an assessment process. Another example is a water and sewer department that uses improved civil engineering models in connection with an improved mapping system to locate and predict waterline breaks. When the reliability of predictions reaches a comfort level for the agency and legislators, then behavior can be shifted from ad hoc repair of water line breaks after they occur to a system whereby actions are taken a priori at appropriate places before breaks occur. If a priori repair is less costly than ad hoc repairs, then the benefits are effectiveness benefits.

Equity benefits are the benefits of making public and private land-use decisions throughout the community with full, informed participation by all interested parties. Increased participation by all citizens is promoted by open access to and use of land data and records with the appropriate ALRS attributes described in chapter 6. Quantitative measures of these benefits are sometimes hard to attain because they are often prospective. Hard does not mean impossible. These are identifiable, important benefits that must be included in a community's assessment of the benefits of investments in modernizing its land records systems. The community's culture of preferred private, local government and participatory land-use governance requires this assessment.

Equity benefits are identifiable and measureable. For example, in Dane County, Wisconsin, the register of deeds created a digital version of the grantor/grantee and tract indexes, created digital images of all newly filed documents, and began working to create digital images of previously filed documents. The register of deeds also invested in the technology and procedures to make all these digital images available electronically, at the cost of reproduction to anyone who wishes to have the material. These improvements allow title insurance companies more efficient access to the title records in the register of deeds. A result of this new technology and the active provision of unfettered access was the unexpected entry of new title insurance companies willing to compete for business with the traditional companies operating in Dane County. This competition lowered title insurance policy costs by about 30 percent, or $300 per transaction per year. The savings to citizens is estimated to be $6 to $7 million annually. The development costs for the register of deeds were less than $500,000

(Tulloch and Niemann 1996). This benefit was unexpected. It turned out to be not only measureable but also considerable.

Unanticipated equity benefits cannot be measured before they occur, which is a problem when a funder demands a measureable benefit/cost calculation. Nevertheless, there is a sense that open, increased access to and use of public land data and records lead to creative and valuable new uses for these materials. The history of GIS/LIS development in America, with its emphasis on openness, is consistent with this sense. Many of the new location-based uses and services for the products of GIS were unforeseen. They were made possible when the investments were made by agencies for the primary purposes of executing their duties. Citizens, groups, and organizations were able to obtain and use the material for other valuable activities.

Increased citizen participation in governance, along with a preference for private and local government land-use decisions, is a cultural goal for many Americans. Increased citizen participation in land-use governance, especially at the local level of government, is an equity benefit served by an ALRS with its emphasis on an information portal. Achieving this equity benefit is also likely to promote continuing support for an ALRS that comes from full community participation in use of system products.

## 7.3.4 AN EXAMPLE: MEASURABLE BENEFITS OF A PRIORI INVESTMENTS IN GEODETIC CONTROL INFORMATION TO SUPPORT ANTICIPATED LAND PLANNING AND MANAGEMENT

Geodetic control information is fundamental material in any land records and information system. This information supports all geospatial activity ranging from the most basic surveying measurements to the sustenance of all that comes from global positioning satellite systems.

The use and value of investments in geodetic control information can be identified, measured, and described. The benefits can be described for the most basic, site-specific land survey. They can also be described in terms of the impact of investments in geodetic control on a community's land planning and management activity. These latter benefits go beyond those for surveyors and geodesists in the normal course of their professional work. For these benefits, the questions become where to look, what to observe, and what meaning to give to the observations.

A first consideration of geodetic control information development focuses on accurate identification, measurement, marking, and recording of the location of points on or near the earth's surface. Considered only in this way, investments in additional geodetic control information involve the search for new technology that reduces the cost of executing the several described actions. The measured benefits of these investments are in the reduced costs of actions using the new methods compared to costs of actions using the old methods. It is noteworthy that the US National Geodetic Survey publishes not just the data but the tools and methods to properly use the data.

However, this approach inspires the question, why invest in geodetic control information at all, and to what extent? Surveyors and geodesists, the primary producers of the information, often respond, "You can't know where you or anything is or make maps without geodetic control." But, how much geodetic control is needed, and how do we know?

The approach can be reformulated. It takes the form of two questions: what unique information is generated by geodetic control location measurements, and how is that information used? Attention shifts from the actions of locating, marking, and recording the locations of points to the full nature and extent of the demand for land records and information made possible by the actions and the measureable use and value of the resulting land records and information.

The information obtained from surveying is the relative location of points on or near the earth's surface. The location, measurement, demarcation, and description actions that constitute surveying can be done without geodetic control information. After all, we must note that Roman surveyors built great projects such as viaducts without geodetic control. They constructed site-specific points that were used to control other measurements.

Geodetic control information is used to establish the relative location of land features and interests in a way that appears to be the same as or similar to that of the Roman surveyors. However, the modern meaning of geodetic control information conveys the sense of actions that extend over a large area and involve a much broader array of actions beyond those associated with a specific site or activity. It is the spatial scale and multiplicity of these sites and activities that characterize the use of modern geodetic activities and the resulting information. This sense of scale is what directs attention to the unique aspect of geodetic control information.

Geodetic control information is the basis for making the relative location of land features and interests compatible with one another. Maps and other products that display or describe relative locations can be generated. Geodetic control information means compatibility of otherwise independent location observations. Universal compatibility is the unique product obtained from the use of geodetic control information. The more land feature and land interest locations made compatible, the greater the sense that a large set of data and records are connected with one another. This attribute of connectedness is a basis for identification of a large array of benefits in geodetic control information beyond that associated with site- and project-specific surveys.

# 7.4   Funding modernization

Efforts to generate funds for modernization of land records, geographic information, and similar systems have typically focused on actions taken at the time records and information are transferred from the agent who creates or holds the materials to those who seek or need them. Efforts to establish and sustain modernization of public land records systems were often based on the concept of recouping investment costs by sales made when data and records were taken out from agencies.

The long debate over the effort to sustain modernization of American public agency records and information systems based on substantial sales of data products continues but may be in its last phases.[3] Two aspects to the debate remain significant. First, several decades of effort to garner sales revenues approaching investment costs have not yielded successful examples.

Second, citizens, groups, organizations, and governments throughout the community—both system developers and users—could benefit from unfettered access to, distribution of, and use of government records and information. Creative, entrepreneurial individuals and organizations have used accessible agency records and information to develop new location information products and services. Widespread access to government databases is not a new idea (Epstein and McLaughlin 1990). Nevertheless, the benefits of widespread database access are now being realized.

System modernization can be supported in ways other than by charging a fee to remove or access material from the system. A major alternative is to charge a fee when records are deposited or introduced into the system. This practice is not new. Document deposition fees have sustained the register of deeds office for a long time.

In some jurisdictions, funding for modernization as well as traditional operation is generated when records are deposited in the system. Modernization efforts sustained in this way extend beyond introduction of new technology to the establishment of new institutional practices and relationships between actors.

The Wisconsin Land Information Program (WLIP, Wis. Stat. §47) is an example. A combination of attorneys, lenders, surveyors, real estate brokers, and other politically powerful professionals, as well as university faculty, convinced members of the Wisconsin legislature to pass land information legislation designed to improve land records systems throughout the state, especially at the county and local levels of government.

Two features of the program recommend themselves to everyone interested in land records and information modernization. One is a provision that requires each county to establish a land information office and a land information officer. The county is given the discretion to establish the operating details for the office and officer. At one end of a range, the county can designate an existing office and officer (e.g., the register of deeds and registrar). The county can delegate advisory powers to the officer in regard to the actions of other land records and information institutions. At the other end of the range, the county can create a new office, a new officer, and an administration with overriding powers that command connections between existing institutions, officials, and records. The choice, and an intermediate option, is available to the county.

A second aspect of the statute is the funding mechanism for modernization. The document filing fee in the register of deeds was increased ($6 per document, originally) to support land information modernization activities. The funding mechanism is based on the long-established concept of a filing fee for records deposition in the register of deeds. This is an a priori funding mechanism. The register of deeds provides a service in the form of a public repository and public notice mechanism for important title documents. The services are paid for when the records are put into the institution, not when they are taken out by those who need them.

In 2013, Act 20, the biennial Wisconsin state budget for fiscal years 2014 and 2015, contained provisions with wide-ranging implications for the WLIP. Changes included the following: (a) an initiative to create an implementation plan for a statewide digital parcel map, the result of a collaborative effort with local governments (Section 186); (b) counties are enabled to retain $8 of the current $30 recordation fee in the register of deeds explicitly for land information purposes (Section 1249). The passage of this legislation in 2013, including the increased funding for land information activities, is remarkable in a period when fees and taxes at all levels of government are subject to close scrutiny by citizens.

The concept of a priori funding is consistent with American attitudes and practices. It can be extended to the realm of land records and information. Those who seek local permits for subdivision, office building, infrastructure, utility plant, or similar developments are often asked to pay a priori impact fees. Applicants for public approval of these developments are asked to pay for increased infrastructure, schools, or other impacts. This widely used concept can include an impact fee for the cost of maintaining the land records and information in a form appropriate for twenty-first century land planning and management.

There are other examples of a priori funding of systems. In Ohio, the county auditor (assessor in other states) operates under a state-enabling statute that diverts a portion of the collected property tax to a fund administered by the auditor to finance office operations. Some auditors have used these funds creatively to introduce new geospatial technology and processes, including databases with some of the aspects of a modern multipurpose ALRS.

# 7.5   The nature of change

The nature and pace of change in land records systems is a function of several forces. If the objective is to modernize the systems so that twenty-first century land planning and management is improved, then two major forces for change can be identified.

One force is the push for change provided by geospatial technology. This technology has driven the exceptional array of new location-based products and services associated with GIS.

However, these new products and services have not always satisfied the demands of all the affected and interested citizens seeking to participate in land planning and management activities to determine the use of land and resources. Those who have the wealth and power, the ability to acquire the data, and the expertise to exercise that power have continued to exert a powerful role in planning and management. They are often reluctant to support changes in the distribution of power that modernization can generate.

The second force for change is the pull of society. This force is based on traditions and practices among members of that society. An example of one of these long-established, preferred traditions and attitudes is the preference for making land-use determinations privately, or by local government if government involvement is necessary. This preference is reflected in the form and function of land records offices such as the register of deeds, assessing offices, and local zoning boards and their symbiotic relationships or partnerships with private professional individuals (e.g., title attorneys, abstractors, and surveyors) and organizations (title insurance companies and property appraisers).

Technology can be used to make the actions of existing public and private organizations more efficient and effective. Its application does not always alter the fundamental relationships between organizations. For example, digitizing indexes and documents in the register of deeds has not altered the fundamental function of the office as a mere repository of documents. The technology can be used to encourage the expansion of the type of documents deposited and actively distributed by the register of deeds.

Alternatively, technology can be deployed in ways that fundamentally alter relationships between organizations. Modernization of existing American land records systems requires changing some of the ways that traditional organizations, public and private, relate to one another to assure that all citizens can more fully participate in land planning and management.

Americans are not monolithic in their attitudes and practices. People in one region may be more disposed to change than people in others. The barrier to change is also affected by which institutions, organizations, and individuals are involved, how they are involved, and who benefits and loses.

Modernization of land records and information based on satisfying the expected records and information demands of all citizens is not the same as modernization based on the supply of materials. Modernization of land records and information systems, represented by an ALRS, can be viewed as a means to alter the balance of power among the several parties interested in land and resource use. This force may not be as dramatically apparent as is the push of technology; however, the slowness of change in land records and information institutions testifies to the strength and importance of existing power arrangements regarding records and information.[4] These forces are depicted in figure 7.3.

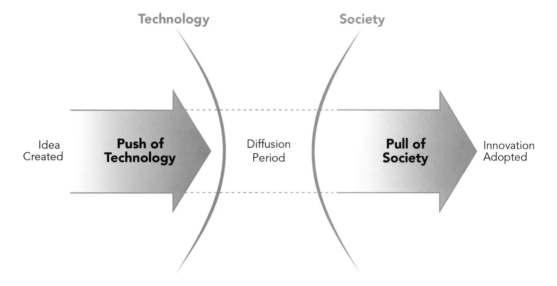

**Figure 7.3** Push of technology and pull of society.

Figure 3.6 courtesy of Bernard J. Niemann. This figure appears in Niemann et al., *Citizen Planners* (Redlands, CA: Esri Press, 2011), 44.

A broad community demand for records and information in support of twenty-first century American land planning and management creates an opportunity for those who design and implement an ALRS. The societal pull represents a force for change to which efforts to modernize the ALRS can be attached. The latent demands for data in service to all citizens in the normal course of all land planning and management activities can be harnessed to serve modernization efforts.

Societal attitudes and practices rarely change in a revolutionary way. The nature, scope, and pace of change are more often evolutionary. Attitudes and practices associated with land rights, interests, and responsibilities are fundamental to land planning and management. Their change is likely to occur in this evolutionary way. The pace of change in American land records and information systems is likely to follow this pace.

Evolutionary modernization of land records systems proceeds through a series of development stages (figure 7.4).[5]

The democratization stage of evolutionary development represents realization of the full set of benefits from modernization of land records systems. These are the benefits

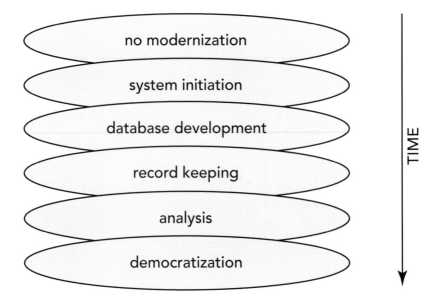

**Figure 7.4** Stages of development.

From Tulloch and Niemann, 1996; courtesy of David L. Tulloch.

of efficiency, effectiveness, and equity that result from both the push of technology and the pull of society. All citizens, groups, organizations, and officials are able to fully participate efficiently, effectively, and equitably in twenty-first century land planning and management.

The challenge is to connect the experts who supply the geospatial technology with those experts who assist the many citizens, groups, organizations, and officials who demand land records and information for land planning and management in accord with American attitudes and practices. The challenge of navigating environmental attitudes is discussed in *Navigating Environmental Attitudes* (Heberlein 2012).

Steinitz has summarized the relationships among actors, professions, sciences, and technologies in the design of the landscape (Steinitz 2012). This is depicted in figure 7.5.

The related roles of professions in this activity are indicated in table 7.1.

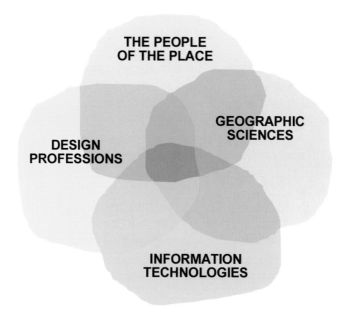

**Figure 7.5** Land planning and management requires the representation of rights, interests, and restrictions for "The People of the Place" when considering the design, planning, and management of land and its resources.

**TABLE 7.1   Land records for the twenty-first century**

|  | **GeoSurvey** | **GeoDesign** | **GeoRegister** | **Land Records Officer** |
|---|---|---|---|---|
| Beliefs | Provide timely and accurate geospatial representations of all public and private land ownership | Provide plans and designs based on how the world works, toward making the world a better place, and redefining our relationship with the environment | Provide open, electronically stored indexes and documents of all rights, restrictions, and responsibilities | Provide the means to collect, share, and maintain the records and information needed to plan, manage, and convey land |

Source: Adapted from Niemann and Niemann, 2013.

## 7.6   Conclusion

The identifiable demands for land records and information by all participants in twenty-first century American land planning are the basis for an ALRS. The ingredients and attributes of an ALRS are described in chapter 6. The demands by many of these participants are not being met by the existing American systems.

The attributes of records and information in existing systems are at the heart of the deficiency. Privately arranged title and property agreements are recorded without significant scrutiny in a public register of deeds, leaving the scrutiny to a long-established process dominated by private actors in a public-private partnership. Science and technology at the beginning of the twentieth century and beyond increased the ability of all actors to gather and use land data and information. The evolution of the land records system necessitated new services, data, and capabilities that were developed by the private sector. At the same time, the large array of public actions that affect use of land and resources in a community are scattered among the many agencies at all levels of government that have land planning and management duties. This condition makes it difficult to assemble a picture, at the time and place when plans, decisions, and actions are taken, of who has the rights, interests, and responsibilities in regard to use of a land parcel. At the same time, geospatial technology generates great quantities of increasingly well-defined data and information about the location of land features. The increasing supply of location-based data and information does not solve the problem of assembling the land records and information demanded by all citizens in law- and legal process–dominated land planning and management.

Changes are needed so that records and information demanded by all participants in twenty-first century American land planning and management are available at the time and place, and in the appropriate form, needed for plans, decisions, and actions. These changes take the form of new actions by traditional land records and information organizations and agencies and new relationships among these actors.

The activities of some agencies and organizations that have begun to make changes indicate that there are large economic and social benefits from investments in these changes. The economic considerations emphasize efficiency, effectiveness, and equity benefits that result from expenditures for system development. These benefits can be identified and measured when considered from a perspective that looks beyond the investing institution to the benefits that accrue throughout the community in the normal course of land planning and management.

A salient feature of these changes is that a new perspective is needed on where to look for the benefits of investments in modernization. In many cases, these benefits are found beyond the agency or organization that invests in modernization. The benefits are in the larger community of citizens affected by or interested in land planning and management. Many current GPS users were not considered when the system was designed and initially built, and many do not care much about the accuracy. Observations indicate that these benefits to the community are large, although they may be small for the individual, unconnected, existing institutions that make the investments within their organization. This perspective on the domain of benefits encourages, perhaps requires, citizens to be involved in the establishment of changes that agencies and organizations make in their land records institutions.

# Notes

1.  The moveable-type printing press is an example of the importance of social factors in the nature and pace of technological development. This press was created in China long before it was known in Western Europe; however, when it did appear in the West during the fifteenth-century Renaissance era, social conditions created a demand for the technology. The demand for books had quadrupled over the previous century; there was no central, continent-wide controlling authority; and geography and meteorological conditions favored local vernacular activity (Boorstin 1983; Diamond 1997).

2. The Urban and Regional Information Systems Association (URISA) was dominated in the 1990s by those who promoted data and information sales as a means by which governments could pay for investments in GIS/LIS. At a URISA meeting in 1989, the president of PlanGraphics, a consulting company on GIS/LIS development, encouraged data sales. In 1993, URISA issued a publication with a set of articles purporting to review the issue of data sales. All but a single, token exception supported sales (Epstein 1993).

3. The effort to secure legislative support for investments in development of public agency geospatial technology, and the data and records products of that technology, has a long history. One argument asserts that investments can be recouped by substantial sales by the agency of the data and records. This assertion is not supported by evidence of substantial levels of data and records sales. A major barrier to an agency holding the data and records for sale comes from the nature of American freedom of information and open records laws. These laws are based on fundamental American attitudes and practices that give citizens access to material used by governments in the normal course of their duties because this information significantly helps citizens know what their governments are doing. Americans prefer the attribute of transparency for their governments. Historian Henry Steele Commager indicated that this preference has a long history when he observed that "the generation that made the nation thought secrecy in government one of the instruments of Old World tyranny and committed itself to the principle that a democracy cannot function unless the people are permitted to know what their government is up to" (Commager 1972, quoted in EPA v. Mink 1973, at 106, the US Supreme Court case interpreting the federal Freedom of Information Act). Access to these materials must come at or near the cost of reproduction, without questions to the citizens about their purposes and without substantial controls on subsequent uses. Despite this history and purpose, some local governments continue to hold data and records tightly and pursue data and records sales.

4. The introduction and diffusion of innovations in a society is a subject of considerable interest. See, for example, Rogers, 2003. For a discussion of this subject in the context of geospatial technology and land planning and management, see Niemann et al., *Citizen Planners* (Esri Press 2010, 44–46).

5. Phased, evolutionary development of both location-oriented GIS/LIS and land records systems with their emphasis on both the nature and extent of land features and interests has been considered (Tulloch and Epstein 2002). The pace of development of GIS/LIS and of land records systems is moderated by the factors of efficiency, effectiveness, and equity.

# Chapter 8
## Actions

The existing American land records and information system is in a state of arrested development. The system inadequately satisfies data and information demands by all those citizens who want to fully participate in twenty-first century American land planning and development. Long-established attitudes and practices regarding land, land records, and land governance keep records disconnected and scattered among many agencies and organizations, and citizens often are unaware of the records' existence. Institutional behaviors and geospatial technology provide both incentives and barriers to the changes necessary for system modernization.

The inadequacies and deficiencies in the American land records and information systems are summarized as follows:

1. Title records that document privately negotiated and arranged transfers of land interests are held and evaluated in a process dominated by a public-private partnership in which fundamental relationships and basic activities have changed little since the nineteenth century. Records of publicly established land interests (e.g., land-use controls) are held separately and organized poorly. The title records system can tell a buyer most of what is known about the allocation of parcel property rights that have been privately negotiated among parties. However, if people and organizations want to know what can and cannot be done with land and its resources, including material concerning the public land-use controls, then the gathering and analysis of records and information is a complex task that is not always completed before plans and decisions are made and actions are taken.

2. Title records and records of public land-use controls are not well connected to data and information about the location of many land features. It is not easy to connect the location of a parcel, natural feature, religious or historical site, and infrastructure elements to the full set of property rights for the feature.

Geospatial technology has dramatically altered the identification, measurement, analysis, and dissemination of data and information about the location of land features and interests. This technology has not been well utilized for the task of connecting records and information about both the nature and extent of land features and interests. It is easier to establish a geographic and land information system (GIS/LIS) filled with data and information about the location of observable land features than it is to fully include the not easily observed rights, restrictions, and responsibilities attached to these features by the community.

3. The public parts of the title records systems are passive institutions that primarily receive, index, and store documents. Emphasis is placed on documents that are important in real estate transactions. Active dissemination and distribution of records by these institutions generally is limited. For example, the register of deeds frequently distributes deposited documents to only the local assessor's office. Those who need an analysis of this real estate information require the service of private professionals who work in a public-private partnership with the register of deeds.

   The demand for land records and information is not limited to material used for real estate transactions. The demand is great and more complex in the realm of the many other actions in twenty-first century American land planning and management. This demand is not well served by the disconnected, passive title records system. Modern information technology has not been deployed to disseminate land records and information to a broader segment of the community and activity. The model of public-private partnerships in the management of title records activity can be extended to this broader array of land planning and management activities.

4. Geospatial technology has satisfied many of the demands for location-based services; however, it has not been fully deployed to satisfy the needs for records and information in the normal course of all American land planning and management. Those who seek a broader array of records and information represent a large and economically, socially, and politically motivated group that can support modernized American land records systems. Their demand for land records and information includes, but is larger and distinct from, the demand for location-based records and information. The demand for records and information from those active in the planning and management process can be connected to the demands of those active in the supply of material by geospatial technology to form a common effort to modernize informational systems and satisfy the needs of all citizens.

## 8.1 Role of an American land records system (ALRS)

An ALRS is specifically designed to provide an antidote to the land records and information uncertainty in American land planning and management. As a process, this system uses preferred American attitudes and practices in the determination of land and resource use. As an object, an ALRS is a repository for a full array of connected records and information about both the nature and extent of land features and interests, which is the material identified for use in the American land planning and management process.

The records and information used for land- and resource-use determinations have the attribute of authority. This material becomes authoritative when it is specifically designated for use in the land planning and management process. This process and its results can be examined to reveal those records and information and their attributes that are given this authoritative status. These records and information become primary candidates for inclusion in an ALRS records and information database. Selection of these designated authoritative records and information, and their prioritized inclusion in an ALRS, imparts an order to the land data and information chaos.

An ALRS is designed to overcome barriers that impede the full deployment of geospatial technology in land planning and management. It provides a means for ameliorating perceived imbalances in the relative strengths of various individuals and groups, legislators, agency officials, and citizens in the land planning and management process by promoting the ability of all citizens to participate in plans, decisions, and actions regarding the use of land and its resources. Full deployment assures the optimal efficiency, effectiveness, and equity in the use of the technology for land- and resource-use determinations.

Achievement of an ALRS is significantly enhanced by development of an open, complete two-way information portal with specific ingredients, such as authoritative records and information. An ALRS would provide mechanisms by which a citizen could not only acquire authoritative records and information but also provide material to receptive public agencies that would actively receive, examine, and where appropriate, give designated authority to the citizen-provided material for use by agencies in the normal course of their duties. This system would be complete in the sense that an ALRS would assemble most if not all of the privately and publicly established authoritative data and records that are or could be used for land planning and management in a community.

An ALRS differs from other geographic, land, resource, environmental, and geospatial information systems in that it *actively* seeks to do the following:

- Overcome barriers to the assembly of a complete record of both the location of land features and the nature of the rights, restrictions, and responsibilities associated with these features.

- Use parcels, parcel maps, and PINs (parcel identification numbers) as bases for connecting records and information about the nature and extent of land interests attached to land features activities of independent land information institutions.

- Develop the concept of a two-way land records and information portal. Provide access by all citizens to open records and information used by agencies in the normal course of their mandated and authorized land-related activities. Encourage agencies to establish administrative procedures that receive citizen-generated records and information, review their appropriateness, and give them authority in the normal course of the agencies' land planning and management.

- Emphasize the goal of a land records and information institution that satisfies the demands of all those who seek to participate in the land planning and management process. The data and information prescribed in or used to satisfy the law and legally prescribed procedures and substantive outcomes of this process can be used as a measure of these demands and a priority for material to be included in an ALRS.

# 8.2   Specific actions

Specific actions needed to realize the concept of an ALRS and achieve its specific objectives include the following:

1. **Assemble, connect, and distribute records and information about privately established land interests.** Privately established land interest records (e.g., deeds) are deposited in the register of deeds. These materials are the basis for activities of other agencies and organizations in the processes that determine the status of title and other land interests. Modern technology makes it efficient, effective, and equitable to expand and improve the connections between the large array of records and information that emphasize the location and status of land features with those in the register of deeds and other offices that emphasize the nature

of land interests. Specific actions designed to achieve this objective include the following:

- Create local government parcel maps in digital format that represent the location of all parcels. Parcels remain central to land planning and management because they are the spatial basis for defining the nature and extent of land rights, restrictions, and responsibilities. Land interests as diverse as scenic easements and notices of building code violations attach to the parcels.

- Generate a unique PIN for each parcel. Standards for PINS have long been a subject for discussion. Although a national system is desirable, it is not necessary. A system is needed that establishes, maintains, and actively distributes unique PINs and requires their use on a variety of land interest documents. The failure to achieve widespread use of PINs to connect location and interest records is a major reason existing systems cannot adequately serve twenty-first century American land planning and management. The local property assessor traditionally develops the community parcel map and generates the PIN. The importance of the assessor in traditional land records activity and the political significance of the office in the local property tax system suggest a continued role for the office in the development of complete, digitized parcel maps and PINs and their connections to other land records and information institutions.

- Expand the number and type of land records and information to which PINs are actively and routinely attached to records and information generated by all agencies in the normal course of parcel-related duties. This action is particularly important in making connections between location-oriented spatial data and records of the nature of land interests. For example, it is not unreasonable in the modern age of geospatial and information technology for the register of deeds to attach a PIN to most documents deposited in the office after reference to the assessor's map, which is now convenient and easy because of modern geospatial technology. Similarly, a building code inspector who identifies a building code violation can attach a PIN to a notice-of-violation document. This document need not only be sent to the violator and posted on the agency website, it can also be deposited in a registry of public actions. The register of deeds may not be the appropriate party responsible for assigning the PIN, but documents deposited in the register of deeds and sent to the assessor can result in a PIN for each parcel in a common index. Modern geospatial and information technology makes it practical to have an active feedback mechanism for the assessor to assign and distribute the PIN to all interested parties. In this context, it is reasonable to require property transferors' and buyers' to have the PIN on the transfer document.

Although PINs are currently used in limited actions, an ALRS is designed to encourage the widespread, if not universal, use of PINs in land planning and management. These actions are often the result of informal, ad hoc arrangements among specific land records organizations. An ALRS encourages widespread use of PINs to connect records and information in most or all institutions. Those who promote the design and implementation of an ALRS can encourage state legislatures, county commissions, and local councils to pass statutes that require this activity among designated agencies.

Geospatial and information technology makes it possible to take these actions efficiently. Benefits in the domain of land planning and management activity make the actions efficient, effective, and equitable throughout the community.

2. **Develop a registry of records and information that documents overriding public interests in land.** A major objective of an ALRS is the development of a registry of public land interest documents.

Documents that record public land planning activities can alter land rights, restrictions, and responsibilities. A major aspect of American land records and information systems is that many, if not most, of the documents cannot be identified, located, and connected to other records because they are scattered among many agencies and without common standards because each agency has its own enabling legislation. These documents do not appear on agency websites, and many people are unaware of their existence. They are needed to complete knowledge of the nature and extent of land features and interests for land planning and management in twenty-first-century America.

Geospatial and information technology makes it convenient and practical to establish a registry of public land interest documents. These documents must be searchable by attributes such as a PIN, location (e.g., latitude/longitude, geographical coordinate), interest, use, and owner. These records include, for example, plats for subdivision approvals, zoning maps, data and maps used to approve permits for new or altered land uses, designated flood hazard maps, and other documentation of public land- and resource-use decisions.

This registry is analogous to the register of deeds for documentation of privately negotiated property interest transfer agreements between parties. The register of deeds does not record the public actions described above. Other terms have been suggested for this public registry, including that of a public property rights infrastructure (Roberge and Kjellson 2009). This registry can be conceived of as a public notice board for public property actions that is consistent with twenty-first-century technology.

This new registry captures the important, authoritative documents that record the array of public actions regarding land use. These can include legislative actions that apply to all land and parcels, such as a proposed or completed statute that describes the details of the process for a permit to construct turbines on a wind energy farm. These documents can also include the details of subsequent actions, such as the geographic coordinates of the locations of constructed turbines.

The registry focuses on these documents among the vast array of material that can now be posted by each person and distributed everywhere. It also focuses on the many records distributed among many agencies at all levels of government. This new registry completes the organization of public records of interests in land and establishes order upon the chaos of land records and information. This record allows for efficient, effective, and equitable determination of what can be done with land. Modern technology makes establishment of this registry practical. Modern twenty-first century American land planning and management requires the creation of such a registry.

The registry of public interest documents in a jurisdiction can begin with a few important public documents and interests and then can be easily expanded. This new registry is facilitated by state and local legislation similar to that for the ubiquitous register of deeds. A new office can be established, or the duties can be assigned to the existing register of deeds, to another existing public office, or to a new office established for the purpose.

3. **Actively define, identify, assemble, organize, and distribute authoritative data used in the documentation of actions implementing overriding public land interests.** Each local jurisdiction can do the following:

   - Identify those public land planning and management activities that have a legally defined demand for specific land data and information. These activities are represented by the processes associated with such things as subdivision permits, zoning variances, wind farm sitings, creation of scenic easements, receipt of documents in the register of deeds, issuance of building code violations, and determination of parcel values for tax purposes. These activities represent significant public actions that, individually and collectively, alter the land rights, restrictions, and responsibilities in a community. The interests are as important as those documented and considered in the traditional land titles system. The full array of interests must be known and considered in twenty-first century land planning and management.

   - Identify those land features for which specific, authoritative land data and information are required by law or that are used in the planning and management process to determine land and resource use. Many of these actions are the result of statutes and rules that describe the specific data and information and their attributes that must be used to execute the action. An example is the locally established ordinance that specifies a Federal Emergency Management Agency (FEMA) flood hazard map as the designated, authoritative information to be used in the determination of properties eligible for low-cost flood insurance. Another example is the legal requirement of the data and information in an environmental impact statement (EIS) prepared to support approval of a permit to drain a wetland. The effect of this action is to designate the material in the EIS as the authoritative data and informational basis for the land-use decision.

- Incorporate and give priority to designated, authoritative data and information in the ALRS database. The data become part of an ALRS database in the way that descriptions in a deed become part of the material used to prepare a parcel map. Authoritative data and information represent a large domain of material that can be used to sustain spatial databases. This material has been identified and distinguished by law and legal process from the vast array of unfiltered material available from Internet sources. Authoritative data still needs professional interpretation. For example, a deed description still requires the work of a professional to interpret its meaning.

- Adopt statutes or administrative rules, standards, and procedures that require or encourage agency officials to actively expand the widespread identification, inclusion, and distribution of authoritative land records and information. This distribution is an expansion and generalization of the long-established practices of the distribution of recorded deeds by the register of deeds to the assessor and the passage of a subdivision permit and plat records and information to the engineer or assessor. Land data and information used authoritatively by one agency can be easily passed to other agencies and to the agency that sustains the local land records and information databases. Geospatial and information technology makes this action practical.

- Adopt statutes or administrative rules and procedures that require or encourage agency officials to actively receive, evaluate, and give authority to citizen-generated and deposited land data and information intended for use in public land planning and management. An example of this activity is the long-established effort of the US National Geodetic Survey. This agency receives citizen observations of conditions that are inconsistent with nautical chart information, examines these observations, and appropriately gives them designated authority by incorporation of the material in the legally binding *Notice to Mariners*. Another example is that of a local agency with the duty to protect the community against invasive species that actively solicits, receives, examines, and gives designated authority to verified citizen observations and then uses the authoritative data as a basis for actions affecting many in the community.

This situation requires the ALRS to establish a defined process wherein appropriate citizen-generated observations are actively distinguished from the vast array of available observations and given designated authority.

4. **Sustain traditional public access to land records and information.** The basis for a modernized land records and information system that serves all citizens in their efforts to fully participate in land planning and management depends on a two-way information portal. An essential element of the system is unfettered citizen access to land records and information used by governments in the normal course of their planning and management activities.

   Open public access to land records and information has been under constant attack, as shown by the continuing efforts by some agencies to withhold data and information to generate revenue from its sale and to assure the confidentiality of some data. ALRS officials must be knowledgeable of the long-established balance between openness and confidentiality that has been established by resolution of debates over the meaning of freedom of information and open records laws. It is appropriate for ALRS officials to be advocates for this traditional balance and its contribution to the ability of citizens to know what their government is doing.

   The actions are summarized in table 8.1.

The action items previously described do not always require new legislation. Many actions can be implemented by administrative rules and procedures consistent with the scope of authority delegated to the agency in the enabling legislation.

Assembly, analysis, and distribution can be achieved with existing authority by use of modern geospatial technology. For example, a register of deeds can expand the nature and extent of the uses for its indexes and records by (1) digitizing indexes and new and old recorded documents and (2) active dissemination of these materials to many organizations, not just the assessor. This kind of activity also occurs on an informal basis when the assessor receives data and information that become part of the GIS/LIS database commonly found in assessors' offices. An ALRS expands this concept and practice throughout the set of community agencies, which expands the set of agencies involved in cooperative reception and distribution. Legislation may be appropriate for institutional stability and for generating activity by reluctant agencies.

**Table 8.1    Actions to establish and sustain an ALRS**

| Action | Records and information | Actors | Initiatives |
|---|---|---|---|
| 1. Assemble, connect, and distribute records and information about privately established land interests | Title documents (e.g., deeds) in the register of deeds, parcel boundary data and maps, etc. | Registrars/clerks, assessors, lenders, title abstractors, attorneys, and insurers | Create, distribute, and attach PINs to maps, data, documents, and other records and information that relate to parcels. |
| 2. Develop a registry of records and information that document overriding public interests in land | Land-use-control legislation, agency administrative actions, court decisions | Legislatures, agencies, courts, executives, etc. | Establish an institution for public land interest records and information within or among existing government institutions. Use PINs and other elements of modern technology to assemble and connect the records. |
| 3. Actively define, identify, assemble, organize, and distribute authoritative data used in the documentation of actions implementing overriding public land interests | Documents that establish and record agency land planning and management actions and contain spatial data and information used to support the agency action | Registrars/clerks, assessors, other government land information agencies | Identify the authoritative spatial data used to support land planning and management and distribute to the central registry of public interest documents. Distribute the data and information to all agencies and organizations that use land records and information. |
| 4. Sustain traditional public access to land records and information | All open land records and information used by governments in the normal course of their duties | All government agencies involved in land planning and management | Sustain traditional open American records systems. |

The number and character of land data and maps designated for use in support of land-use decisions can be expanded. Existing examples of this activity include the designated nautical charts for pilot decisions, designated flood hazard maps for identification of the location of buildings eligible for low-cost flood hazard insurance, and designated plats for subdivision approval. In some jurisdictions, specific wetland maps can be designated for use in support of local administrative actions, such as road expansion.

An engineering-based, informal dispute-resolution activity can be established in a public office without legislation. Officials can evaluate boundary evidence presented by all parties to a boundary dispute. Resolution of the dispute includes designation of data and information as the basis for the dispute resolution. The parties to the dispute can agree to be bound by the resolution. Deposition of the agreement in the register of deeds or the registry of public land interest documents gives additional authority to the agreement and widespread notice to the community.

Formal connections between existing, independent land records institutions can be expanded. It is common, for example, for the register of deeds and subdivision-approval agencies to transfer data and records to an assessor.

Although no technical or economic reason exists to restrain the nature and scope of these new activities and relations among land records institutions, long-standing, institutional barriers need to be broken down. Benefits to the community are incentives for change in the independent organizations.

# 8.3   Conclusion

An ALRS with concepts and practices described in this book establishes an institution that actively assembles and widely distributes the land records and information sought by all those interested in twenty-first century American land planning and management. An ALRS seeks to overcome the barriers that make it difficult and expensive to acquire a reasonably complete and clear picture of both the nature and extent of land features and interests and to do so at the time and venue when land- and resource-use determinations are made.

Existing land records and information institutions continue to keep separate data and information about the location and physical status of land features from records that describe the allocated rights, interests, and responsibilities associated with these features.

Existing institutions also keep separate records and information about privately arranged and asserted land interests, such as those commonly described in title records, from records of publicly established land interests, such as those that appear in land-use controls.

Existing institutions favor ad hoc efforts to assemble a picture of the status of land features and their associated land interests.

Existing institutions continue to encourage public agencies to withhold land records and information from easy public access despite freedom of information and open records laws that encourage their widespread distribution consistent with the goal of providing ways for citizens to know what their government is doing.

A major aspect of an ALRS is its constant attention to the demands for appropriate land data and information by all parties interested in twenty-first century land planning and management. Land planning and management are dominated by laws and legal processes that reflect preferred American attitudes and practices regarding land, land records, and land governance. These laws and processes can be identified and examined for important, detailed observations and messages about the demands for land records and information, and their attributes, that are used to determine land and resource use. These demands can be synergistically combined with the supply of land data and information and its distribution provided by geospatial and information technology to generate widespread support for land records and information system modernization.

Design and implementation of an ALRS continues to rely on the long-established role of private-sector actors and organizations who provide ad hoc analyses of the status of land features and titles as well as assurances of the results of these analyses. An ALRS retains the option of a system of government assurance of the status of titles and boundaries as it continues to rely on traditional methods and actors in the land records modernization effort.

Many, if not most, of the changes needed in processes and products for development of an ALRS are well within the nature and scope of existing legal regimes and geospatial technology. The problems associated with long-established human relations within and between institutions have stunted the development of American land records and information systems.

The demand for land records and information by all participants in twenty-first century American land planning and management is a force for change. Measureable benefits of a modernized system are found throughout the community. The benefits from an institution's investments in land records modernization often are in a realm beyond that institution's traditional activity.

These conditions suggest that the development of a local, modernized system, an ALRS, requires those who operate existing land records and information systems to look beyond their traditional activity and to form partnerships with the many who seek material in order to better participate in land planning and management. Failing to do so will limit full deployment of geospatial technology and, more importantly, negatively affect the quality of twenty-first century American land planning and management determinations.

# Glossary

**authoritative data** Data designated by statute, administrative rule, court opinion, or recognized common practice as a basis for determining the use of land and its resources. These designations in the public land planning and management process influence the data used in private land planning and management. These data acquire a level of authority when they are actually used in a process defined by law and process.

**boundary** The spatial limit of land interests, or property rights, associated with a parcel. A parcel boundary map represents the intended location of a boundary. The location of objects placed on the ground in consequence of the intention to create a boundary at the place indicated on a map or in a location description constitute the primary evidence of the boundary location according to American law.

**cadastre** A record of interests in land, encompassing *both* the nature and extent of these interests. The definition used in the seminal, modern American work on the subject, *Need for a Multipurpose Cadastre*, prepared by a panel assembled by the American National Research Council and published by the National Academy Press in 1980. A cadastre often contains information about the value of the parcel and other data and information useful for land planning and management. In many countries, cadastres also include the legal status, or authority, given by law to the cadastral record of ownership and boundary.

**cadastral map** A map showing the locations and boundaries of recognized land features and interests associated with a parcel. Other data and information may be placed or connected to the parcels represented on the map.

**cadastral surveying** In the United States, this term usually refers to the use of modern land measurement science and technology to measure, represent, and record observations of land features that indicate the location and extent of land interests, especially the surface boundaries of parcels.

**chain of title** The set of deeds and other legal instruments prepared over time that document transfers of land interests between buyers and sellers in a deeds-based conveyancing and recording system such as exists in the United States. The system in the United States differs from that commonly associated with a cadastre in other nations wherein the status of ownership revealed by the chain of title is assured by a government. In the United States, private professionals and organizations locate, summarize, interpret, and assure the status of the documents in the chain of title and the status of ownership revealed by the contents of the documents.

**chaos** This is the result of the ability, made possible by modern spatial and information technology, for each person to place any data anywhere. Access to these data is very desirable, especially for location-based services. However, the widespread availability of all kinds and conditions of data is a problem because of the varying and often uncertain quality of the data. This is an acute problem where standards of law and legal process determine the specific data and information and their attributes, which is required for use in the normal course of land planning and management by agencies and organizations. Chaos is reduced when land records systems include processes that identify this specified, required data and give the data priority in design and implementation of the system.

**conveyance** The transfer of land rights, interests, and responsibilities from one owner to another. The transfer may be a full set of privately held rights or of an individual right, such as a mortgage, mineral right, or easement.

**deed** A legal document that describes both the nature of land interests conveyed and the boundaries of these interests.

**deeds recordation** A tracking system that records land interest conveyance documents in a public office. In the United States, this is usually the local government's register of deeds. The office makes no statement about the validity of assertions and evidence of interests described in the documents.

**demarcation** The marking of boundaries of land parcels on the ground.

**easement** An interest held by one party in land held by another.

**geodetic control information** Data and information resulting from a network of monuments on or near the earth's surface whose locations are determined according to high standards for observations and using advanced land measurement technologies. Geodetic control information is the basis for global positioning satellite (GPS) systems.

**geographic information** The common reference to data and information about the location and extent of land features. It is also common to use the term to refer to such data when the material is in a digital form for storage and use in a computer. However, the term can be used to refer to land data and information in any form, analog or digital.

**geographic information system (GIS)** An integrated collection of computer software and data used to view and manage information about geographic places, analyze spatial relationships, and model spatial processes. A GIS provides a framework for gathering and organizing spatial data and related information so that it can be displayed and analyzed.

**Global Positioning System (GPS)** A system of radio-emitting and receiving satellites used for determining positions on the earth. The orbiting satellites transmit signals that allow a GPS receiver anywhere on earth to calculate its own location through triangulation. Developed and operated by the US Department of Defense, the system is used in navigation, mapping, surveying, and other applications in which precise positioning is necessary.

**land administration system** An infrastructure for implementing land planning and management policies and strategies. The infrastructure includes institutions, a legal framework, processes, standards, data and information, management, and dissemination activities.

**land governance** The activities and processes associated with the roles of citizens, groups, organizations, and agencies in determining and implementing plans, decisions, and actions concerning use of land and its resources.

**land information management** The management of all kinds of data and information about land features and interests. Both the nature and location of these features and interests are included.

**land information system** A name often used interchangeably with a geographic information system (GIS).

**land interests** Rights, restrictions, and responsibilities that individuals and communities attach to land features and allocate to individuals and groups. The features can be natural, such as rivers, wetlands, flood hazard areas, and soil. The features can also be man-made, such as buildings, parcels, and historical sites. Recognition of land interests can be informal or formal, as occurs when they are the result of a law or legal process.

**land planning and management** Related public and private processes (plans, design, decisions, and actions) that determine the use of land and its resources in accord with preferred attitudes and practices in a community.

**land records and information** Documents that record both the nature and extent of land features and interests. They record the results of public and private land planning and management activities that affect land interests, land use, and the land data and information that support these activities. Land records and information are the physical manifestation of this activity and may be in analog or digital form.

**land records system** The institutions and processes that collect, manage, and distribute records of land features and interests about both the nature and extent of these features and interests. A land records system, such as an American Land Records System (ALRS), is similar to a cadastre without the requirement of government assurance of the ownership status that the records indicate.

**land registration** A process of recordation of land transfer documents and legal recognition by a government agency of the status of land interest ownership that results from the transfer.

**land tenure** The manner or system of holding land interests (rights, restrictions, and responsibilities). Examples include feudal, freehold, and communist systems.

**land transfer** The transfer of land rights, restrictions, and responsibilities.

**metes and bounds** A description of the boundaries to a parcel that refers to the bearings and lengths of the boundary lines (metes) between boundary marks on the ground, together with reference to the names of adjoining parcels (bounds).

**monumentation** The process of placing objects on the ground to identify land parcel boundaries.

**multipurpose cadastre** A cadastre that contains not only records and information about both the nature and extent of land interests, but also data and information about a variety of land features needed for a wide variety of land planning and management activities. A cadastre that is designed to emphasize the need for data and information that serves single or limited purpose, such as a fiscal/tax or a title/ownership purpose, is not a multipurpose cadastre. It is common in America to find land records and information systems designed and operated to serve one or a few uses (e.g., register of deeds, assessor). The products of these single- or limited-purpose systems are sometimes used for purposes other than those for which they were designed and by people or groups beyond the originating organization, which can have undesirable results.

**overriding public land interest** An appropriate exercise of the government's power to tax, take, and control private land interests. Examples are found in the set of public land-use and zoning controls that limit the exercise of land-use determinations that can be made by the owner of private interests. These overriding public land interests usually are documented in the records systems of the many agencies that exercise and implement them. However, there are no central registries in American jurisdictions that assemble documentation of these important land interests, such as a registry analogous to the register of deeds for private land interest documents.

**parcel** An area of land within which there is a recognized set of land rights, restrictions, and responsibilities.

**parcel identification number (PIN)** A unique reference that identifies a parcel, usually a number. A PIN can be attached or linked to parcel maps and parcel location data. It can also be attached to textual documents that contain information about the nature or substance of land features and interests. The use of PINs to link a wide variety of records of both the nature and extent of land features and interests is essential for the development of a modern land records and information system that satisfies the information demands in land planning and management.

**property** An object capable of being owned by an individual or group that has a legally recognized, allocated power to determine how the object is used.

**spatial data and information** Data and information about the location of a land feature or interest on or near the earth's surface.

**tenure (or land tenure)** The manner in a community whereby people hold and exercise land rights, restrictions, and responsibilities.

**title insurance** A system, common in the United States, wherein assurance of the status of ownership of land interests in a parcel is provided by a private organization, not the government.

**Torrens, or Torrens-type, cadastre** A cadastre wherein the status of ownership of land interests is assured by a government agency.

**trust** A land interest held by one party for the benefit of another party. It is a land interest transferred by one party to a trustee with the intent that the trustee manages the interest for the benefit of another party.

# References

Administrative Procedures Act (APA). 1946, as amended. 5 USC § 500 *et seq.*, Pub. L. 79-504.

Andrews, R. N. L., ed. 1979. *Land in America: Commodity or Natural Resource?* Lexington, MA: Lexington Books, DC Heath.

Antennuci, J. 1989. A declaration opening the presentation of a paper at the annual meeting of the Urban and Regional Information Systems Association (URISA), Boston.

Babcock, R. F., and D. A. Feurer. 1979. "Land as a Commodity Affected with a Public Interest," In *Land in America: Commodity or Natural Resource?*, edited by R. N. L. Andrews. Lexington, MA: Lexington Books, DC Heath.

Barlowe, R. 1978. *Land Resource Economics: The Economics of Real Estate,* 3rd ed. Englewood Cliffs, NJ: Prentice Hall.

Black's. 1981. *Black's Law Dictionary.* St. Paul, MN: West Publishing.

Boorstin, D. 1983. *The Discoverers.* New York: Random House.

Bradford, W. 1948. *The History of Plymouth Colony.* Roslyn, NY: Walter J. Black.

Brown, C. M., W. G. Robillard, and D. A. Wilson. 1986. *Boundary Control and Legal Principles,* 3rd ed. New York: Wiley.

———. 1981. *Evidence and Procedures for Boundary Location,* 2nd ed. New York: Wiley.

Clapp, J., J. McLaughlin, J. Sullivan, and A. Vonderohe. 1989. "Toward a Method for the Evaluation of Multipurpose Land Information Systems." *URISA Journal* 1: 39–45.

Clawson, M. 1972. *America's Land and Its Uses.* Baltimore: RFF Press.

Clean Air Act (CAA). 1970, as amended. 42 USC § 7401 *et seq.*, Pub. L. 91-604.

Clean Water Act (CWA). 1972. Federal Water Pollution Control Act (FWPCA), as amended.

Coase, R. H. 1960. "The Problem of Social Cost." *Journal of Law and Economics* 3: 1–44.

Commager, H. S. 1972. Quoted in Environmental Protection Agency (EPA) v. Mink, 410 US 73, 93 S. Ct. 827. Quoted at 93 S. Ct. 845 from *The New York Review of Books,* October 5, 1972, p. 7.

Cowen, D. 1994. "The Importance of GIS for the Average Person." In *GIS in Government: The Federal Perspective, Proceedings of the First Federal Geographic Technology Conference,* September 26–28, Washington, DC, 7–11.

Cronon, W. 1983. *Changes in the Land.* New York: Hill and Wang.

Daylor, R. F. 1982. "Land Information in the Permit Process." In *Proceedings of the International Symposium on Land Information at the Local Level*, edited by A. Leick, Surveying Engineering, University of Maine, Orono, 89–99.

Deetz, J., and P. S. Deetz. 2000. *The Times of Their Lives: Life, Love, and Death in Plymouth Colony.* New York: W.H. Freeman.

De Steiguer, J. E. 1997. *The Age of Environmentalism*. New York: McGraw Hill.

Diamond, J. M. 1997. *Guns, Germs, and Steel*. New York: W.W. Norton.

Dodd-Frank Act (Dodd-Frank Wall Street Reform and Consumer Protection Act). 2010. Pub. L. 111-203.

Duchesneau, T. D. 1982. "Determining the Economics of Ground Control Based Upon a Geodetic Reference System." In *Proceedings of the International Symposium on Land Information at the Local Level*, edited by A. Leick, Surveying Engineering, University of Maine, Orono, 219–23.

Ellickson, R. C., and A. D. Tarlock. 1981. *Land Use Controls: Cases and Materials.* Boston: Little, Brown.

EPA v. Mink. 1973. 410 US 73, 93 S. Ct. 827, 35 L.Ed.2d 119 (1973).

Epstein, E. F. 1993. "The Case Against Commercialization of Public Information." In *Marketing Government Information: Issues and Guidelines.* Urban and Regional Information Systems Association (URISA). Washington, DC, July 1993, 101–03.

Epstein, E. F., and P. Brown. 1989. "Land Interests." In *Multipurpose Land Information Systems: The Guidebook,* 2 vols. Reston, VA: Subcommittee on Geodetic Control.

Epstein, E. F., and T. D. Duchesneau. 1984. "Use and Value of a Geodetic Reference System." Federal Geodetic Control Committee (FGCC), NOAA, Department of Commerce, HD 108.6.E7. April 1984. See *URISA Journal* 2(1), 1990, for a shorter version.

Epstein, E. F., and J. D. McLaughlin. 1990. "A Discussion on Public Information." *ACSM Bulletin* 128: 33–38, American Congress on Surveying and Mapping, Washington, DC.

Euclid v. Ambler. 1926. 272 US 365.

Federal Emergency Management Act (FEMA). 1974, as amended. See 6 USC § 313.

Federal Freedom of Information Act (FOIA). 1966, as amended. See 5 USC § 552(b).

Federal Geographic Data Committee (FGDC). 2008. "Authority and Authoritative Sources: Clarification of Terms and Concepts for Cadastral Data," Version 1.1. August 2008. FGDC Subcommittee for Cadastral Data. See also FGDC Subcommittee for Cadastral Data, Cadastral NSDI Reference Document, http://nationalcad.org/download/cadastral-nsdi-reference-document.

Federal Water Pollution Control Act (FWPCA). 1972, as amended. FWPCA § 404. See 33 USCA §§ 1251–1387. This statute is commonly known as the Clean Water Act (CWA).

Fiorino, D. J. 1995. *Making Environmental Policy.* Berkeley, CA: University of California Press.

Foresman, T. W., ed. 1997. *The History of Geographic Information Systems: Perspectives from the Pioneers.* Upper Saddle River, NJ: Prentice Hall.

Fox, Jr., W. F. 1986. *Understanding Administrative Law.* New York: Mathew Bender.

Freedom of Information Act (FOIA). 1966, as amended. 5 USC § 552 *et seq.*, Pub. L. 89-554.

Friedman, L. M. 1985. *A History of American Law,* 2nd ed. New York: Touchstone.

Getches, D. H. 2009. *Water Law in a Nutshell,* 4th ed. St. Paul, MN: Thomson West.

Hallam, E. M. 1986. *Domesday Book through Nine Centuries.* London: Thames and Hudson.

Hardin, G. 1968. "The Tragedy of the Commons." *Science* 162: 1243–248.

Heberlein, T. A. 2012. *Navigating Environmental Attitudes.* New York: Oxford University Press.

Heclo, H. 2008. *On Thinking Institutionally.* Boulder, CO: Paradigm Publishers.

Henry, N. 1975. *Public Administration and Public Affairs.* Englewood Cliffs, NJ: Prentice Hall.

Hibbard, B. H. 1965. *A History of the Public Land Policies.* Madison: University of Wisconsin Press.

Hurst, J. W. 1956. *Law and the Conditions of Freedom in the Nineteenth Century United States.* Madison: University of Wisconsin Press.

———. 1975. Lectures notes from the course "American Legal History," Law School, University of Wisconsin–Madison. Autumn 1975.

Johnson, H. B. 1976. *Order upon the Land: The US Rectangular Land Survey and the Upper Mississippi Country.* London: Oxford University Press.

Kishor, P., B. Nieman, D. Moyer, S. Ventura, and P. Thum. 1990. "Lessons from CONSOIL." *Wisconsin Land Information Newsletter* 6: 1–13.

Koontz v. St. Johns River Water Management District. 2013. US Supreme Court, No. 11-14475. Argued January 15, 2013. Decided June 25, 2013.

"Land Ordinance of 1785." *Journals of Continental Congress* (May 20, 1785) 28: 375. Library of Congress.

Lang, L. 1995. "The Democratization of GIS." *GIS World* 8: 62–67.

Leopold, A. 1949. *A Sand County Almanac, and Sketches Here and There.* New York: Oxford University Press.

Mackaay, E. 1982. *Economics of Information and Law.* Boston: Kluwer-Nijhoff.

McLaughlin, J. D. 1975. "The Nature, Design, and Development of Multi-Purpose Cadastres." PhD diss., University of Wisconsin–Madison.

Mladenoff, D. 2009. *Wisconsin Department of Natural Resources Magazine.*

See http://www.dnr.wi.gov.wnrmag/2009/68/insert.pdf.

MERS. 2013. For an introduction to MERS, see the MERS website: http://www.mersinc.org. See also the *New York Times* MERS information page: http://topics.nytimes.com/top/news/business/companies/mortgage_electronic_registration_systems_inc/index.html.

MOLDS. 1975. *Proceedings of the North American Conference on Modernization of Land Data Systems.* Washington, DC, April 14–17. Library of Congress Catalog Number 75-18651.

———. 1979. *Land Data Systems Now: Proceedings of the Second MOLDS Conference.* Washington, DC, Oct. 5–7. Library of Congress Catalog Number 79-64891.

Mill, J. S. 1993. *On Liberty and Utilitarianism*. New York: Bantam Books.

Morgan, E. S. 2006. *The Puritan Dilemma: The Story of John Winthrop*. New York: Pearson Longman.

Moyer, D. 1975. *Problems of Land Ownership Data and Related Records in Data Needs and Data Gathering for Areas of Critical Environmental Concern: Part 2*. Madison: Institute for Environmental Studies, University of Wisconsin–Madison.

———. 1980. "Property, Information, and Economics: A Foundation for Land Information System Evaluation." *Geo-Processing Journal* 1: 275–95.

National Environmental Policy Act (NEPA). 1969, as amended. NEPA § 101(c). See 42 USC § 4321, *et seq.*, Pub. L. 91-190.

National Research Council. 1980. *Need for a Multipurpose Cadastre*. Washington, DC: National Academy Press.

———. 1982. *Modernization of the Public Land Survey System*. Washington, DC: National Academy Press.

———. 1983. *Procedures and Standards for a Multipurpose Cadastre*. Washington, DC: National Academy Press.

———. 1993. *Toward a Coordinated Spatial Data Infrastructure*. Washington, DC: National Academy Press.

———. 2007. *National Land Parcel Data: A Vision for the Future*. Washington, DC: National Academies Press.

Nautical Charts. http://www.navcen.uscg.gov/. See also A. L. Shalowitz. *Shore and Sea Boundaries*. Washington, D.C.: US Department of Commerce, National Oceanic and Atmospheric Administration. 1962 (vol. 1) and 1964 (vol. 2).

New Brunswick. 2012. *Land Registry – Land Titles Search/Registration http://app.infoaa.7700.gnb.ca/gnb/Pub/EServices/ListServiceDetails.asp?ServiceID1=1775&ReportType1=ALL*

Niemann, Jr., B. J., D. D. Moyer, S. J. Ventura, R. E. Chenoweth, and D. A. Wiskowiak. 2010. *Citizen Planners: Shaping Communities with Spatial Tools*. Redlands, CA: Esri Press.

Niemann, Jr., B. J., and S. Niemann. 1994. "GIS Innovator: Innovation with Affect—Part Two." *GeoInfo Systems* (September) 4(9): 46–52.

———. 2013. "Navigating Land Modernization Attitudes: Sustaining Wisconsin's Land Information Modernization Journey." Twenty-sixth Annual Meeting of the Wisconsin Land Information Association, Lake Geneva, WI.

Ostrom, E. 1990. Governing the Commons: The Evolution of Institutions for Collective Action. Cambridge, UK: Cambridge University Press. See also "Building a Better Micro-Foundation for Institutional Analysis." 2005. *Behavioral and Brain Sciences* 28(6): 831–32.

Pennsylvania Coal Company v. Mahon. 1922. 260 US 392.

Philbrick, N. 2006. *Mayflower: A Story of Courage, Community, and War*. New York: Viking (Penguin).

Pinto, J., and H. Onsrud. 1995. "Sharing Geographic Information Across Organizational Boundaries: A Research Framework." In *Proceedings of the 1995 URISA Conference*, San Antonio, Texas, 688–94.

Rapanos v. US. 2006. 547 US 715.

Real Estate Settlements Procedures Act (RESPA). 1974, as amended. 27 USC §§ 2601–22617.

Real Estate Settlements Procedures Act (RESPA). 1978. Booz Allen Hamilton, Inc., RESPA reports prepared for HUD (Draft), Washington, DC. This report was not made final and published.

Roberge, D., and B. Kjellson. 2009. "What Have Americans Paid (and Maybe the Rest of the World) for Not Having a Public Property Rights Infrastructure?" International Federation of Surveyors. www.fig.net/pub/fig2009/papers/ts04a/ts04a_roberge_kjellson_3287.pdf.

Rogers, E. M. 2003. *Diffusion of Innovations, 5th ed. New York: Free Press*.

*Shelley v. Kraemer. 1948. 334 US 1.*

Shick, B. C., and I. H. Plotkin. 1978. *Torrens in the United States: A Legal and Economic History and Analysis of American Land-Registration Systems*. Lexington, MA: Lexington Books DC Heath.

Sierra Club v. S. C. (County of Orange). 2013. 57 Cal 4th 157.

Smiley, J. 1991. *A Thousand Acres*. New York: Fawcett Columbine.

Smith, Z. A. 2009. *The Environmental Policy Paradox, 5th ed. Upper Saddle River, NJ: Pearson Prentice Hall.*

Steinitz, C. 2012. *A Framework for Geodesign: Changing Geography by Design*. Redlands, CA: Esri Press.

Stensvaag, J-M. 1999. *Materials on Environmental Law*. St. Paul, MN: West Publishing.

Tarlock, A. D., J. N. Corbridge, Jr., and D. H. Getches. 2002. *Water Resource Management: A Casebook in Law and Public Policy, 5th ed. New York: Foundation Press.*

Tocqueville, Alexis de. 1945. *Democracy in America*. Volume 2. Translated by Henry Reeve. Revised by Francis Bowen. Edited by Phillips Bradley. New York: A. A. Knopf.

Tulloch, D. L., and E. F. Epstein. 2002. "Benefits of Community MPLIS: Efficiency, Effectiveness, and Equity." *Transactions in GIS* 6(2): 195–212.

Tulloch, D. L., and B. J. Niemann Jr. 1996. "Evaluating Innovation: The Wisconsin land Information Program." *GeoInfo Systems* 6(10): 40–44.

Urban and Regional Information Systems Association. 1993. *Marketing Government Geographic Information: Issues and Guidelines*. Washington, DC: Urban and Regional Information Systems Association.

United States Constitution, Article 1, §3, §§18.

Vonderohe, A. P., R. F. Gurda, S. J. Ventura, and P. G. Thum. 1991. *Introduction to Local Land Information Systems for Wisconsin's Future*. Madison: Wisconsin State Cartographer's Office.

von Meyer, N. 2004. *GIS and Land Records: The ArcGIS Parcel Data Model*. Redlands, CA: Esri Press.

Webb, W. P. 1959. *The Great Plains*. Lincoln: University of Nebraska Press.

Williamson, I., S. Enemark, J. Wallace, and A. Rajabifard. 2010. *Land Administration for Sustainable Development*. Redlands, CA: Esri Press.

WLIP (Wisconsin Land Information Program). Wis. Stat. § 47.

Worster, D. 1985. *Rivers of Empire: Water, Aridity, and the Growth of the American West*. New York: Pantheon.

White, C. Albert. 1983. *A History of the Rectangular Survey System*. Washington, DC: Bureau of Land Management (BLM), US Department of Interior.

Zinn v. State. 1983. 112 Wis 2nd 417, 334 N.W. 2nd 67.

# Index